CHALLENGES AND OPPORTUNITIES

In Exponential Times

CHALLENGES AND OPPORTUNITIES
IN EXPONENTIAL TIMES

Insights, Innovations, & Trends

Z. S. Andrew Demirdjian, Ph.D.

To order additional copies of this book, contact:
Xlibris
1-888-795-4274
www.Xlibris.com
Orders@Xlibris.com
542250

Contents

PART II: *Basic Challenges in the Business World*

This Book is Dedicated
to

Ms. Zara Mokatsian whose provocative ideas
outside the box have always entertained me
intellectually and have stimulated my thinking
beyond the ordinary for some years.

ENDORSEMENTS FOR CHALLENGES AND OPPORTUNITIES IN EXPONENTIAL TIMES:
INSIGHTS, INNOVATIONS, & TRENDS

"Dr. Demirdjian's books are remarkable for their clarity, scholarship, and uninhibited expression of unusual and futuristic concepts. This book will keep you abreast of the things, concepts, and theories to come in the next decade and beyond."

Thang Nguyen, Ph.D.

"Lucid explanations of most important concepts of the last 250 years. An essential aid to understanding both modern history of science and technology including contemporary social, political, and cultural developments. An enlightening book to read."

Robert Chi, Ph.D.

"From reverse evolution to realism, from democracy to dialectics, *Challenges and Opportunities in Exponential Times* is an essential expose of important concepts of recent years. The book is written in clear, simple language; it makes even most complex ideas such as genomics, epigenics or the theory of psychological evolution immediately comprehensible."

David Bojarsky, Ph.D.

"Professor Demirdjian's new book again presents all major challenges and opportunities that one can foresee in a rapidly changing world. It deals with challenges in business, opportunities in emerging technologies and fundamental issues facing our society. Through numerous examples of recent experiences, the author reinforces his thoughts and ideas so readers, students and executives alike, could understand his points. I recommend all to read this book."

M. B. Khan, Ph.D.

"Sections on art, science, technology, philosophy, politics, and the "human sciences" of psychology, anthropology, Sociology and economics provide a comprehensive excavation of the modern mind in the form of ideas which shaped the world as we know it and will continue its invasive course into the future generations to come. A provocative book, to say the least"

Steven V. Le, Ph.D.

PREFACE

S CIENCE IS FICKLE like a pretty mercurial woman. It has the tendency to change from time to time. I believe a personality trait like that makes science interesting and worth pursuing.

Before 1880s scientists thought the world to consist of five continents. Then came Alfred Lothar Wegener to change that belief. This German scientist was the first to suggest that about 250 million years ago, there was one huge supercontinent known as Pangaea. Before this time, the real "beginning" is still shrouded in mystery.

As we can see, Wegener's concept has caused a paradigm shift in scientific thinking about our world. The truth about science is both revolutionary and evolutionary. So, the chapters in this book contain the present knowledge if not truth about various sciences.

For example, nanotechnology is hailed as life-saving science, perhaps like DDT and asbestos, it can be disowned for causing the world more problems in the health of plants and animals than anticipated previously.

Therefore, we should read these chapters with an open mind and refrain from becoming dogmatic about their nature and potential.

Meanwhile, when different sciences morph, it is advisable that we get acquainted with the insights, innovations, and trends of scientific thinking in various fields. The main purpose is to help guide us in our efforts to move forward in business and social interactions.

As Arthur Compton (1892-1962), the American physicist and Nobel Laureate has once stated succinctly that the benefits of science are not only material ones. The truths that science teaches us are of common interest the world over. Science, like music, has no national boundaries. The language of science is universal, and is a powerful force in bringing the peoples of the world closer together.

The ideas presented in these chapters would help policy decision makers all over the world choose viable courses of action based on the current scientific principles. Equipped with avant-garde ideas, one would be able to navigate better through the mystery of life on the planet Earth.

Undoubtedly, we live in exponential times. The urgency of it all does cause one to pause and think about what the future holds. Due to new scientific discoveries, on account of new concepts, things are changing so rapidly in the landscape of life.

In fact, changes are constant in our personal and professional lives, and manufacturers need to consider the ways the world is changing in order to be relevant with the right products, processes, and service to capitalize on the future.

That knowledge is exploding is beyond any debate. Top management have a tremendous challenge to manage their own information and intellectual property, let alone be able to access and leverage the information available across the globe.

Search, Knowledge Management (KM), and Business Intelligence (BI) will become bigger requirements inside PLM (product life cycle management) and to drive product innovation by tapping into global knowledge sources.

Social computing will also play a role here, as manufacturers try to discover the people with the right knowledge in addition to knowledge.

Time to market new products, services, and social and political ideas is evaporating. The time lag between a technical advance and the commercialization is disappearing. This makes new product development (NPD) critical, but also further supports the need to rapidly discover and take advantage of knowledge anywhere in the world. It also means that top management will have to get their products right the first time, or someone else will take the market away from them.

Computing power is exploding beyond our imagination. The exponential growth of computing power will play a large role in, for example, what PLM vendors are able to do with their software, opening up new opportunities including continued expansion of 3D, animation, and simulation in the way we interact with products.

In my first book titled **Challenges and Opportunities in Changing Times** I had twenty chapters. In this book I have thirty-eight chapters. The additional eighteen chapters provide a wider range of topics on recent science and technology innovations, making it a rich source of information for the student as well as for the corporate executive.

In reviewing related innovations, trends and perspectives, I have presented some insights on the times we live in and my thoughts on the implications for new and old science and technology, and on new and old social concepts.

When we look at some of the trends, they will illustrate the fact that we live in exponential times. I hope you will find them interesting to read and reflect on them.

Therefore, each chapter has the potential to help the reader, be it a student or a professional, understand the nature, limitations, and the prospects of emerging science and technology, the parameters of paradigm shifts in social and economic areas in exponential times.

By exponential times, it is figuratively meant that our society is growing in exponential leaps and bounds by virtue of new ideas. Hence, there are things we must realize to prepare for the future, to own our tomorrows. As John Galsworthy has warned us aptly: "If you do not think about the future, you cannot have one."

Z. S. Andrew Demirdjian, Ph.D.

ACKNOWLEDGEMENTS

THE ORIGINAL MODEL in Chapter Twenty-Five, "Immigration Behavior: Toward a Social-Psychological Model for Research," is dedicated to my editor, Ms. Zara Mokatsian, for her readiness to suggest novel ideas.

If we gotten out of the Stone Age, it is not because we had run out of stones, but rather humankind had discovered other technologies to use.

Source Unknown

INTRODUCTION

Challenges and Opportunities:

The Two Sides of the Same Coin

MOST OFTEN WE dissociate challenges from opportunities as though these two words are basically different. For example, if your friends were to invite you to join them on an expedition to Mount Ararat that may immediately be seen by some negatively as a difficult task to climb that sacred and lofty mountain.

With a slight shift in perception, however, you may see the task positively. How exciting it would be to climb the Biblical mountain, to see the surrounding valleys from the top, to tell or even brag to your friends and family members about this once-in-a-life-time experience.

How we represent things to ourselves determines how we will respond to any given situation. In turn, our response will help determine the outcome.

Challenge is most often associated with negativity (e.g., with difficulty), while an opportunity is represented positively as a way to achieve one's goals. In fact, an opportunity may prove to be as difficult as a challenge. For example, suppose you were given a hefty scholarship (i.e., the opportunity) to further your studies at college. To achieve and maintain the required GPA (grade point average) would be a big challenge (a difficult task to achieve).

When life hands you a challenge, what would be your initial response on an emotional level? Some of us will see the challenge as involving difficult tasks, while, others would see it as an opportunity to get ahead in life.

How we respond to any challenge or opportunity reveals a lot about our attitude and perception. Some of us will see challenges as opportunities. It means that they have a healthy degree of optimism, self-confidence, open-mindedness, and a lively spirit for adventure. It would show that we enjoy life and look forward to tackling whatever comes next. Some of us will adhere to the old adage: nothing ventured, nothing gained.

On the other hand, what does it say about us when we greet a new challenge with an expression like: "Oh no, I don't know how I can cope with this." We immediately turn blind to the opportunities in the given challenge.

Confronted with challenges and opportunities, some of us fail to see the two sides of the same coin. If two things are two sides of the same coin, they are very closely related although they seem different.

On the other hand, some of us see challenges and opportunities as being poles apart. This would indicate that some of us are running low on resources. It reveals a pessimistic personality, closed minded and somewhat fearful perception of the world around us. Such a limiting attitude can only attract more of the same. Fear and negativity deprives one of a life of joy and prosperity.

Let us heed what Sir Winston Churchill had to say about our perception of challenges and opportunities: "The pessimist sees difficulty in every opportunity. The optimist sees the opportunity in every difficulty."

The surprising thing is that there is often only a small degree of difference between a positive, optimistic perception, and a negative, pessimistic one. Even though these two attitudes are polar opposites, they both started with the same challenge.

Let us suppose you say: I am one of those who see a challenge negatively and an opportunity positively. Such a statement should not make you an island in an alien archipelago, sometimes inaccessible by leaps, flights, and voyages on vessels of human behavior. There are many people like you who fail to discover opportunities hidden in a challenging situation.

Here is a story for you to see how to convert a challenge into an opportunity:

Once a king of ancient times wanted to know how his people would react when faced with a challenge (or obstacle). Therefore, the king placed a huge boulder in the middle of a busy roadway. Then, he hid himself and wanted to see if anyone would remove the huge rock. Some of the king's wealthiest merchants and courtiers came by and simply walked around it. Many loudly blamed the king for not keeping the road clear, but none of them did anything about getting the stone out of the way.

Then a peasant came along carrying a load of fruits and vegetables. Upon approaching the rock, the peasant laid down his burden and tried to move the stone to the side of the road. After much pushing and shoving, he finally succeeded to roll the boulder off the road. After overcoming the challenge, the peasant picked his load of fruits and vegetables, he noticed a fat purse lying on the road where the huge rock had been. The purse contained many gold coins and a note from the king indicating that the gold was for the person who removed the boulder from the roadway.

This story tells us that the peasant noticed the challenge (obstacle, i.e. the huge rock), he did not walk simply around the obstacle, he abstained from blaming the king, he did not run away from the obstacle, but pushed the rock to the side of the road after so many efforts and finally got the gold coins. This story could be taken as a classical example of how a challenge can provide an opportunity to improve one's lot.

Here are some ways one can turn challenges into opportunities:

One way is to accept responsibility by liberating yourself. The first thing to do is to recognize that we are in control. Furthermore, we need to accept responsibility for our responses, and recognize that they assert a powerful influence on our life.

When confronted by a challenge, we cannot just "say I see nothing but difficulties in that situation. I am just a pessimist." Until we accept responsibility we will not have any reason to change. Accepting responsibility is a wonderfully liberating experience. It puts us in control of our own life. That means that we are in the driver's seat and can change directions at will. This type of feeling empowers oneself quite well.

Some people avoid responsibility because it brings with it accountability. Let us ask the following: Is it more empowering to be accountable for one's own actions and attitudes, or to make somebody else responsible. The danger lies in the fact that when we give away accountability, we create a state of helplessness. Therefore, we need to liberate ourselves by accepting responsibility.

A second way to perceive a challenge as containing opportunities in it is to use leverage. Leverage means that you exert the greatest amount of control with the least amount of effort. Timing is everything in this situation. When one is confronted with a challenge, the first thing to do is to avoid negative thinking for it is much more difficult to reverse the course later on. Once the first step is controlled, one can start out in the right direction to perceive a challenge as an opportunity. Such an attitude would make it much easier to maintain that positive direction.

Life is full of temptations. If one likes exotic places for vacation, such as Hawaii, and is financially strapped, instead of yielding to the desire, one would employ leverage in controlling the wish to go to Hawaii. Instead, he would be content to spend his vacation in his backyard. In the same vein, weighing a situation before we respond to it gives us the greatest leverage in determining the outcome. In other words, get the negative perception out of your mind when confronted with a challenge.

A third approach to enable us to perceive a challenge as an opportunity for advancement, would be to remind oneself of positive outcomes accrued from tackling a challenge in the past. In this way, remembering the good things that came out of facing a challenge would boost one's confidence, self-esteem, and outlook on life.

A fourth way is to list down on a paper or take a mental note of all the major opportunities inherent in a challenge. We are living through challenging times, but inherent in nearly all challenges are hidden opportunities for renewal and transformation.

A fifth approach would be to grin and bear it when confronted with a challenge rather than escape from it by burying your head in the sand. The determination to face a challenge squarely is half the battle.

A sixth way would be the fear of failure. When we analyze our attitudes, we realize that the biggest problem to overcome is our fear. This is the most damaging part of our lives. It simply blinds us to see opportunities in a challenge.

For example, a person would like to learn how to ride a motorcycle. He may know theoretically all the rules, but the greatest challenge is to jump on the bike and learn how to ride it. The biggest challenge is fear. Fear of falling off and hurting oneself since compared to an automobile an accident proves to be more fatal when one is riding a bike than while driving a car. Fear leads to many negative questions in our minds. If I fall off the bike, I may break a leg. Then I won't be able to work for months. In this way, he would psych himself out of learning how to ride a bike. Fear prevents us from facing a challenge and from seeing it as an opportunity. Only through action a challenge can be faced and end up being an opportunity.

The seventh way to convert a challenge into an opportunity is to equip oneself with knowledge and experience about the requirements inhering in a certain challenge. In today's Cyber Age, one hardly needs to go to the library for getting some information about an issue. We have the luxury of the Internet right in our homes. Before prejudging a challenge as to be ignored or avoided, read about it, explore its positive outcomes, weigh the tasks involved in it, and determine the opportunities gained from dealing with a certain challenge.

People have different attitudes and responses when they are faced with a challenge or obstacle in their daily lives. Many of us are afraid of seeing and facing challenges. Many of us simply talk away the challenge by dwelling on its negative possible outcomes. Many of us keep on blaming the structures, the powers to be, policies, rulers, etc. to justify to ourselves for not facing a challenge ourselves. Running away from the challenge or obstacle would make us fail to realize that every challenge in our lives provides us an opportunity to improve our lot in life and living conditions.

So let us cultivate our entrepreneurial spirits and see the world through positive lenses. Here is what has been said succinctly about this long before we had begun to explore the relationship between challenges and opportunities:

"Entrepreneurs are simply those who understand that there is little difference between obstacle and opportunity and are able to turn both to their advantage." Niccolo Machiavelli (1469-1527).

Z. S. Andrew Demirdjian, Ph.D.

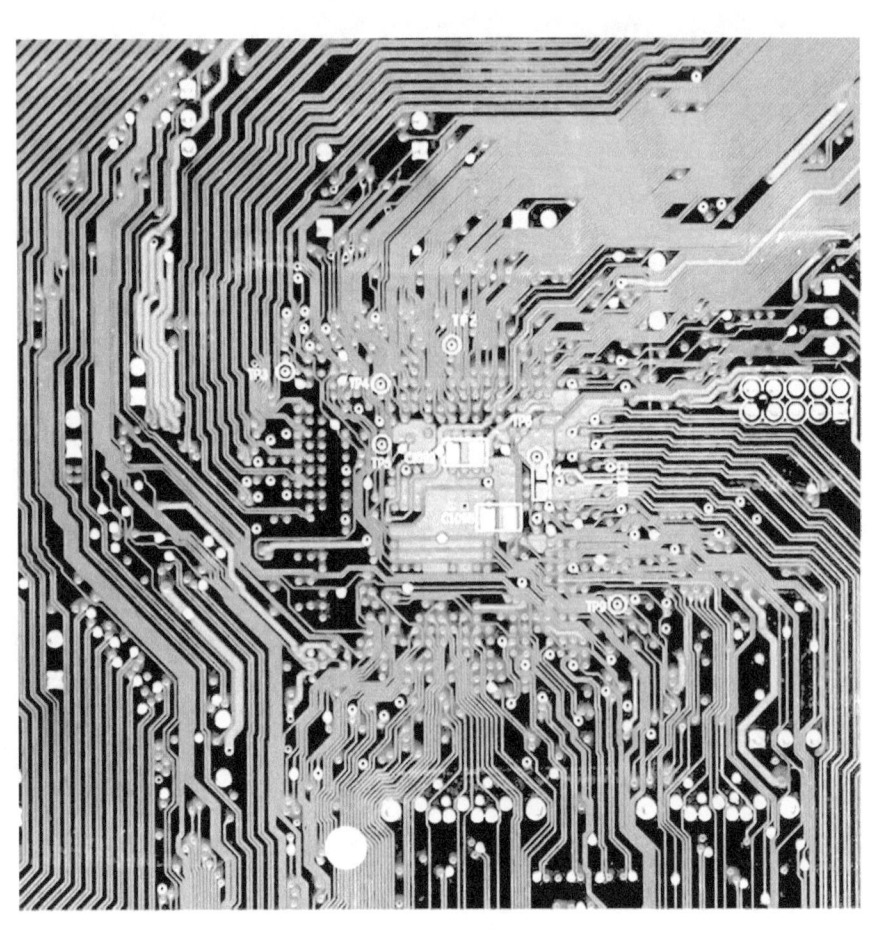

Part I

Challenges and Opportunities in Emerging Technologies

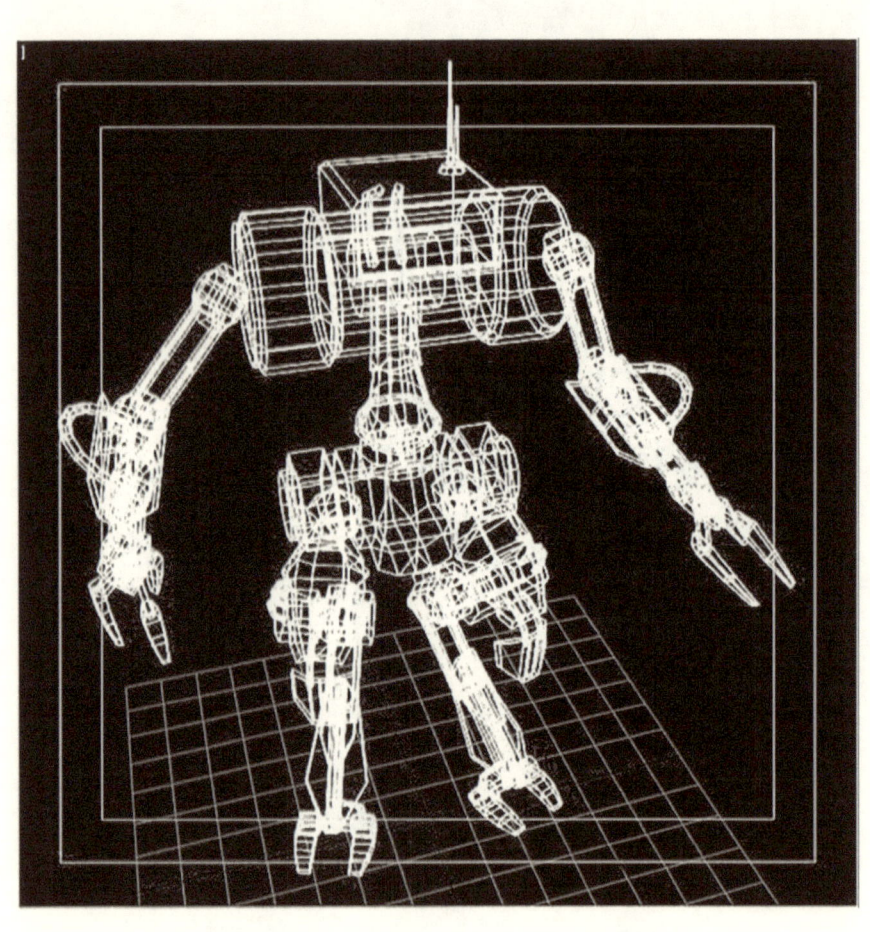

CHAPTER ONE

An "HRP-4" in the White House!

UNLIKE THE ECONOMIST Thomas Malthus's (1766-1834) dismal prediction that the world population would increase exponentially to the point when humanity would starve, I happen to subscribe to that school of thought that human ingenuity would always bail us out by coming up with new technologies to provide us with the necessities and even luxuries of life.

I have always looked forward to the day when science would perfect a robot to act precisely like a human being. Recently, I was reading that Japanese researchers had invented a robot that in the near future could replace humans in the workplace.

The last statement kindled my interest to read about this prospect further which has immense implications particularly to business and industry and to society in general. In this way, the business sector would be able to control the cost of labor and society would benefit from "in-sourcing" rather than the current outsourcing dilemma.

The Ultimate Robot

The machine, the robot, was invented by specialists from Kawada Industries and the National Institute of Advanced Industrial Science and Technology (AIST). The team entertains the lofty hope that with their latest invention they would take a giant step towards developing a robot that could solve the problems of labor shortage in the land of the rising sun.

Mr. Noriyuki Kanehira, robotic systems manager at Kawada, told reporters that his team of scientists had developed a working machine "in the image of a lean but well-muscled track-and-field athlete." This statement was made at a news conference in Japan where the researchers presented their blue-and-white robot classified or named "HRP-4". The number "4" in the name signifies the fourth robot invented by the research team at Kawada.

The robot presented at the conference measured 152 centimeters (59 inches) tall. During this public demonstration, the robot had stood on one foot, twisted its waist, struck poses, walked according to the given voice commands and had tracked different objects by moving its head.

It is interesting to mention that the machine supposedly moves more freely than the previous models of its kind. According to the inventors, this robot (HRP-4) is able to run a series of separately-developed software applications, reports, and so on.

Commercialization of the HRP-4 Robot

Here is the most interesting part: researchers had begun selling their latest invention in the beginning of this year. Currently, the machine will be mainly sold to universities and research institutes in Japan and aboard. The price tag for this low cost model is around 26 million yen (306,000 dollars). The developers of this machine are expecting to sell three to five units a year.

I do not know about you, but I would like to see a robot invented to perform like a human being unlike our politicians. Had we had a robot in the White House who would not lie, deceive, or manipulate the public, we would not have been dragged into protracted, costly wars such as in Iraq and Afghanistan, and now the President is threatening to go to war in Syria. If robots could replace humans, it would be great for they would not have forked tongue and be programmed to do things for the benefit of society.

On the other hand, I am afraid that robots would also replace office workers, executives, janitors, etc. We would end up jobless for sure. Despite all that deprivation, most people would like to see an HRP-4 running the nation based on honesty and integrity.

CHAPTER TWO

The Race for the Nanotechnology Leadership is Revving Up

D AVID HUME (1711-1776) once emphasized about the anthropology of working together as a team by stating that "Everyone has observed how much more dogs are animated when they hunt in a pack, than when they pursue their game apart." In recent years, the world has witnessed how China has taken giant steps to pursue nanotechnology for military and civilian purposes.

China Aspires to Be the World Leader

Presently, China is entertaining great expectations to the extent of becoming the world leader in the science and technology of this emerging field. Such exuberance is predicated on the fact that, like dogs hunting in packs, China has formed a formidable alliance consisting of the government, the business sector, and the academia to synergize the time and the talents of the tripartite.

By observing and studying China's application of anthropological concepts such as core group theory and practice, federated groups, and group dynamics in business and industry, one would learn a lot from their experiences because this once "sleeping giant" is now fast becoming the quarterback in this far-reaching scientific competitive game known as nanotechnology.

Basically in industry, nanotechnology is the art of manipulating materials on a very small scale in order to build microscopic machinery. Nan(n)os is a Greek noun for "dwarf, little old man." The present meaning of nano, of course, stands for a symbol representing something extremely small, one billionth

Nanotechnology is very diverse, ranging from extensions of conventional *device physics* to completely new approaches based upon *molecular self-assembly*, from developing *new materials* with dimensions on the nanoscale to investigating whether we can *directly control matter on the atomic scale.*

Nanotechnology product categories currently run the gamut of cancer cures, medical diagnostics, energy transformation panels, silicon replacement, enhanced coating, efficient printing, space travel, and defense technology to cite a few.

The Drive to Research for New Nanoproducts

Currently, the focus has been on discovering new materials, novel phenomena, new characterization tools, and fabricating nanodevices by the integration of engineering, science, and biology. Like past innovations, the new technology has problems even though its prospects look bright. While nanotech shows great promise, we should heed the warnings that it also has many perils hidden as unintended consequences or as plain side effects on humans and as well as on the environment.

As the promising opportunities are becoming more evident, the industrial nations are hopping on the bandwagon to benefit from nanotechnology. None of the superpowers are wider awake about the exponential potential of nanotechnology than the People's Republic of China.

Napoleon Bonaparte once said of China, "Let her sleep, for when she wakes up, she will shake the world." The French Emperor's clairvoyance was on target. Today, the sleeping giant is the rising nation and her people and products are affecting the world as never before.

That China is a large country is an understatement. It has a mega land, it has a mega population, and it enjoys a mega capital. It has so much surplus money that it frequently loans money to the United States government. With a teaming population of 1.3 billion highly motivated citizens, the industrial nations are courting China for its mega market too.

While the average Chinese has improved his standard of living, there are still millions and millions of people in China who have to subsist on less than a few dollars a day. Will there be better days for those at the bottom of the economic totem pole? All indicators point in that direction with confidence. One major reason is that China is revving up the nanotechnology revolution by forming an alliance among three powers: the government, the business sector, and the academia.

Progress is seldom achieved without a proper organizing effort. The Chinese officials have recognized the necessity of forming an alliance to expand the horizons of nanotechnology in their own country rather than remain copy cats and pirate the copy rights of someone else's inventions.

Before the turn of the century, the Chinese media had neither mentioned nanotechnology nor had it covered its promise to revolutionize China's high tech industry. Despite that oversight, today China has many research centers and many enterprises engaged in the production of technologies in a multibillion dollar industry.

Among the most prominent research centers are in Beijing, Shenyang, Shanghai, Hangzhou, and Hong Kong. Collectively, they account for over 90 percent of all nanotech R&D. The drive of these centers is to improve China's nanotech R&D and commercialize nanomaterials, nanoelectronic components, nanobiological and medical technologies.

China Forming a Federated Group for R&D

By drawing upon recent business anthropological concepts of core group theory and practice, a nation can achieve wonders in research and development. Joining together into a federated group is the key to success. China must have adhered to the preceding concepts; it formed an alliance consisting of the core groups in the government, the business sector, and the academia.

These various core groups have synergized into a federated working machine: effective organizations need effective core groups, and in federated organizations or alliances, where there are dozens more core groups, there is a chance for a far greater effectiveness than among the core groups of a single company.

One of the major players in the Chinese nanotech alliance is the Chinese government. Ever since the National 836 Hi-Tech R&D plan, huge investments for nanotech projects poured from both the central and local governments. This kind of state funding was designed to transform China's nanotech industry by 2010, putting the nanotechnology on par with China's microelectronics, telecom, and other Hi-Tech industries. Furthermore, China is applying the anthropological concept of working together (i.e., team work) for better results.

While the United States' pursuits of new technologies are on individualistic basis, China believes in group effort, concerted effort, in mobilizing all available resources for greater effectiveness. This contention will be clear, when we next discuss as how China is fast becoming the premiere country, the flag bearer in the new arena of nanotechnology. The China's forte seems to have been based on its focus on federated core groups.

The second major player in China's rapid advances in nanotechnology is the business sector. For example, in 2001, Shanghai Nanotech Promotion Center was established to focus on R&D and the industrialization of tools needed for nanotech research.

China also enjoys the advantage in research of nanometer materials. By the time the center opened, China had more than 300 enterprises in the nanotech sector, with some 7,000 scientists engaged in nanotech R&D.

The third major player in the Chinese nanotechnology revolution is the academia. The Nanotechnology Partnership between the Chinese Academy of Sciences (CAS) and Veeco Instruments is one big example. In 2002, CAS launched a joint project with the U.S. company, Veeco Instruments, Inc. The CAS Institute of

Chemistry and Veeco formed the partnership to co-operate in the running of a nanometer technology center aimed at providing access to Veeco-made nanotech instruments to Chinese researchers.

The main reason as to why Veeco chose to invest in nanotechnology projects in China was based on the confidence of the superiority of Chinese nanotechnology. The partnership between CAS and Veeco came amidst great optimism regarding China's nanotech potential. Veeco executives had been certain that China would gain the leadership position in nanotech. This bold statement of confidence in Chinese nanotech superiority has been affirmed by CAS executives who knew that China enjoyed the advantage in research of nanometer materials.

The Outlook for Nanotechnology for China

The outlook for nanotechnology R&D in China is not rosy, it is not promising, it is simply explosive. According to the Scientific American, by 2015 products incorporating nanotech will contribute approximately $1 trillion to the global economy. In terms of world employment, over two million workers will be employed in nanotech industries, and three times that many will have supporting staff jobs.

When it comes to China, though, one thing is certain for sure: the nanotechnology mobilization will, if it has not already, undoubtedly launch China as the undisputed leader in the world. Currently, several dozens of institutions are engaged in basic nanotech research.

If Napoleon Bonaparte were to visit China today, he would dub the once sleeping giant's quantum leaps in nanotech nothing short of the "Miracle of the 21st Century!"

CHAPTER THREE

Trends in Exponential Times

THROUGHOUT HISTORY, INVENTIVE minds have imagined and then created new things to make their lives better, easier, and more enjoyable. Man did not get out of Stone Age because he ran out of stones. His large brain made him find other materials such as iron to make stronger implements out of it. From the wheel to penicillin to the computer, inventions continue to change the way we live and work. From financial standpoint, business and industry have been the greatest beneficiaries of most inventions.

As to who came up first with the idea of the wheel is lost in antiquity, but the Sumerians in ancient Mesopotamia may have invented the wheel and the Assyrians exploited it by using two-wheeled chariots in warfare. The result was the building of one of the greatest empires of the ancient world. Today, the wheel has transformed humankind in many ways. For example, without the automobile, society would revert to crawling in moving around for business or pleasure.

In modern times, Penicillin was discovered in 1928, after Alexander Fleming accidentally left a dish of staphylococcus bacteria uncovered for a few days. He returned to find the dish dotted with bacterial growth, apart from one area where a patch of mold (Penicillin notatum) was growing. The mold produced a substance, named Penicillin by Fleming, which inhibited bacterial growth and was later found to be effective against a wide range of harmful bacteria. When Penicillin was finally isolated by other researchers, it became the wonder drug for the development of lifesaving antibiotics to treat mankind against many infections and deadly diseases.

Great inventions in the past were milestones separated from each other in time by many years. Nowadays, the rate of speed of inventions is highly accelerated since we currently live in exponential times. In the Cyber Age of rapid technological innovations, harbinger of things to come are of keen interest to those who would like to keep abreast of future changes.

Some Examples of Future Inventions

Here are some examples of the anticipated inventions and trends based on emerging science and technology that are destined to impact the way we do things:

➤ Hairy Surface. Engineering researchers headed by Professor Wolfgang Sigmund at the University of Florida have invented a new plastic surface by imitating the minute hairs that grow on the bodies of spiders. Unlike the existing products, the artificial hairy surface is water-repelling surface which uses no chemical treatments. The surface refuses to get wet and it is almost perfectly hydrophobic.

Any possible product use for Hairy Surface? Any one with elasticity of the mind would say "sky is the limit." One can take a copy of the Playboy magazine while snorkeling in a coral sea!

➤ Hands Free Texting. Under Dr. Juan Gilbert's supervision, researchers at Clemson University in South Carolina have developed a hands-free alternative to cell phone texting while driving. The application is called Voice TEXT that allows drivers to speak text messages and keep their eyes on the road at the same time. This invention is music to the ears of especially female drivers who enjoy more ear "gratification" while driving than their male counterparts.

➤ Office Paper into Toilet Paper. A machine was invented by Oriental Co. Ltd of Japan, dubbed as the "White Goat," which transforms office paper into toilet paper. Reportedly, the machine uses 40 sheets of standard A4 office paper and water to make one roll of toilet paper in roughly 30 minutes. The machine is slated to be sold for $100,000 which will be out of the reach of most households. This is an improvement over our cave ancestors who used stone to wipe themselves off after you know what. It is one of the latest innovations coming out of the eco movement. Trees of the world will rejoice when less of their brothers and sisters will be slaughtered to satisfy humankind's insatiable appetite for paper.

> Organic Transistors. The French National Science Agency, under the nanotechnology research director Dominique Vuillaume, have developed a hybrid nano-particle-organic transistor that can mimic the main functionalities of a synapse. This transistor is named the NOMFET (Nanoparticle Organic Memory Field-Effect Transistor) an organic device made of a molecule called pentacene and gold nano-particles. It functions similar to a biological spiking synapse and can pave the way to a new generation of neuro-inspired computers responding to external stimuli in a manner similar to the nervous system. Most likely, one may in the near future see computers for sale with a sticker that reads "NOMEFT Inside" rather than the usual "Intel Inside."

➤ Microring Wireless Devices. Most of us are familiar with wire clutter at home and at work. Soon the intertwined bundle of wires will be a thing of the past. At the Purdue University, researchers have developed a system capable of converting ultra-fast laser pulses into bursts of radio-frequency signals applying "Microring resonators." Microring resonators are tiny silicon devices consisting of a radio transmitter made into a solid-state device. The tiny circuits are activated by 100 femtosecond (100 quadrillionths of a second) bursts of light from diode lasers. This kind of technology will be useful for high-definition television broadcasts, Internet connections, and even for transmitting wireless signals inside cars. Microring Wireless would be a movable feast for those mobile professionals who "hate" rivers of tangled wires and who are constantly on the go.

➤ Microsoft's Pictionaire. Despite recent advances in tablet-computing and touch screens, loss is still great in transferring ideas from brain to hand to a glass screen with a stylus. To circumvent this limitation, Microsoft and UC Berkeley researchers have developed an interactive tablet called "Pictionaire" which extends surface computing to support collaborative design teams by allowing for the digitization of real world objects. Pictionaire works by making digital copies of physical objects put on the screen with overhead digital camera, allowing a user to manipulate real objects with all the fun touch screen interaction can provide. The result is the augmentation of surface computing for real world objects. Anyone interested in producing a sequel to the "Avatar" movie, should have one of those state-of-the-art Tablet PCs with an interactive table called Pictionaire.

➢ Seeing Sound. Synesthesia is a condition in which one type of stimulation evokes the sensation of another, as when the hearing of a sound produces the visualization of a color. So far Synesthesia functionality depended on the imagination of the individual. For centuries, scientists have been attempting to make sound and vibration visible by way of exciting media like liquids and particles. This field is now recognized and the Swiss medical doctor and pioneer, Hans Jenny, named it Cymatics (from Greek: kupa "wave") which is the study of visible sound and vibration, typically on the surface of a plate, diaphragm, or membrane. Cymatics is also known as "modal phenomena." Directly visualizing vibrations involves using sound to excite media often in the form of particles, pastes, and liquids. The possible applications of Cymatics are wide-reaching; it is being applied in fields as diverse as oceanography and sound healing. Other researchers are experimenting to uncover the nature of sound to simultaneously please the eyes as well. In other words, make sound serve as "candy" to the eyes. As science and technology advances, it has the tendency to deepen the ignorance of the masses. Only mavens, opinion leaders, geeks, and the like embrace new inventions readily and thus drive a wider gap between themselves and the majority of people who are technologically challenged bystanders.

Innovations Are Double-Edged Swords

For instance, some twenty years ago preparing an article for publication was less complicated.

The researcher would write the article and then turn it over to the competent department secretary (Excuse me, now they are called Administrative Support Coordinators) to type it. Nowadays, the researcher has to write the article, to word process it, and finally format it according to some journal guidelines. In other words, the researcher has to be well versed in the computer application technology (i.e., Word Perfect or Microsoft Word) which is constantly evolving and leaving the average person fall behind in the new technology savvies. The result is fear and frustration when it comes to constant change brought about by new equipments, methods, and applications.

While one should appreciate the fruits of science and technology, one should also remember that innovations have the double-edge sword tendency to create complications for the majority of people and widen the generational gap. Therefore, from human perspectives of any invention, one should always keep in mind that for some people new inventions would complicate life rather than improve it, at least in the short run, until one gains some confidence and competence to know how to apply the new inventions.

CHAPTER FOUR

Nanotechnology:
Hopes for Limitless Horizons

I MAGINE A WORLD teaming with billions of desktop-size machines that can create almost anything useful to humankind just in minutes – clothing, furniture, electronics, medical devices, cosmetics to mention a few. Presently, such devices are not available yet, but one day soon a small nano-factory sitting on your kitchen counter would let you order nearly anything you need or want at little or no cost.

One would command: "Computer, make me breakfast, home fries, orange juice, and coffee." This is a dream come true for many bachelor men. Although such prognostications would sound like something out of science fiction, according to futurist Ray Krzell, nano-factories could one day soon be providing an individual and his or her family with meals, medicines, and most essentials no later than mid 2020s.

In layman's terms, nanotechnology refers to materials, applications and processes designed to work on extremely tiny scales (e.g., to give you a perspective, a nanometer is one-billionth of a meter). Many unique properties and uses can be derived from structures built at the nanoscale, giving this growing technology an enormous potential for future development.

A Brief History of Nanotechnology

Nanotechnology, as an emerging field of science, was first alluded to in 1959, but remained largely theoretical until the 1980s. The invention of the scanning tunneling microscope

(STM) allowed the first direct manipulation of individual atoms. Furthermore, an outstanding breakthrough happened in 1989 when IBM used such a machine to spell out their corporate logo – using just 35 atoms!

Arthur C. Clarke once observed by saying that "Any sufficiently advanced technology is virtually indistinguishable from magic." The most recent magic for humanity is nanotechnology.

In a matter of thirty years, various structures were developed, each built on an atom-by-atom basis. Presently, nanotechnology is among the fastest growing areas of science and technology, with exponentially fast progress being made.

Recent Breakthroughs in Nanotechnology

Just some of the recent breakthroughs include:

*New fabrics that are highly resistant to liquid, causing it to simply fall off without leaving any dampness or stains.

*The first integrated circuits using three-dimensional carbon nanotubes. These could be vital in maintaining the growth of computer power, allowing Moore's Law to continue.

*Military equipment made lighter and stronger through the use of nanomaterial composites.

*Solar panels with greater efficiency through the use of nanotechnology materials.

*Water purification bottles, with filters only 15 nanometers in width, allowing military personnel and also civilians hit by disasters to create safe drinking water (even if that water comes from a filthy source).

*Nanostructured polymers in display technologies allowing brighter images, lighter weight, less power consumption and wider viewing angles.

*Nanotechnology surfaces which are highly resistant to bacteria, dirt and scratches.

*Nanostructured catalysts used to make chemical manufacturing processes more efficient, saving energy and reducing waste products.

*Pharmaceutical products reformulated with nanosized particles to improve their absorption and make them easier to administer.

There are many other applications and contributions to science and technology. The list is growing all the time. By 2025, nanotechnology is expected to be a mature industry, with countless industrial and consumer products.

The Phenomenal Growth of Nanotechnology

The graph below charts the phenomenal growth of nanotechnology since 1970s and like Duracell batteries, it will continue to grow strongly.

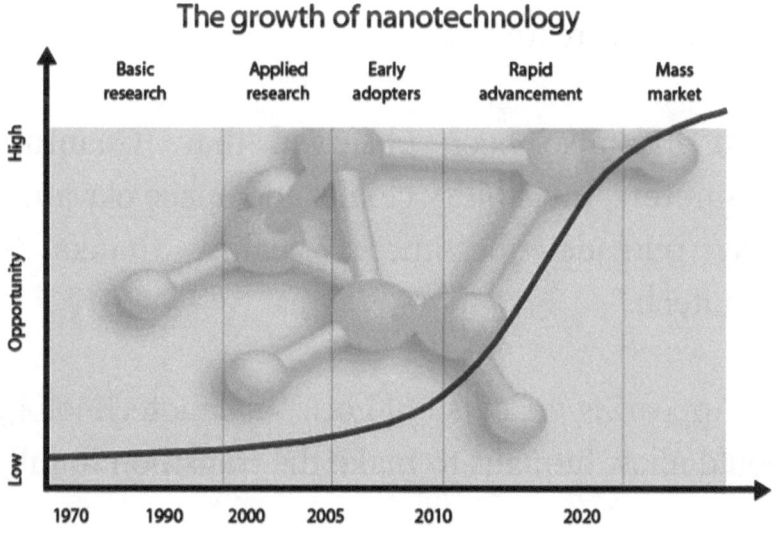

The growth of nanotechnology

Considering the future horizons, nanotechnology will play a major role in medicine and longevity. For instance, blood cell-sized devices will go directly into the human body, eradicating pathogens and keeping people healthy. Full-immersion virtual reality and other advanced concepts will become possible through the use of these "nanobots".

Looking into the future, so-called "nanofabricators" would allow the creation of macro-scale objects on an atom-by-atom basis. For example, home appliances using this technology could serve as 3-D printers – downloading products from the web and literally building them from scratch. Physical items would each have their own code or algorithm that would program the machine to create them.

It has also been observed that "Quantum computers, invisibility cloaks and space elevators may one day become a reality, thanks to nanotech."

Appearing on a distant horizon, nanotechnology could allow humans to make the transition to fully non-biological forms. Entire bodies and brains could be reconstructed at the atomic scale, leading to practical immortality. That is what most people would like to have in order to beat the cruel "father time" which like a river waits for no man on its course to age anything in its path.

Concerns over Toxicity of Nanomaterials

Like most new sciences, the implications of nanotechnology are beset with heated debate and controversy. As social and behavioral scientists claim, nothing in nature is either totally good or totally bad. Nanotechnology is no exception.

Nanotechnology has the potential to produce radically new materials and devices with a vast range of applications in such areas as engineering, medicine, electronics and energy production. However, concerns abound about the toxicity and environmental impact of nanomaterials, and their potential harm effects on global economics including speculations about various doomsday scenarios.

The above-mentioned concerns have led to a debate among advocacy groups and governments on whether special regulation of nanotechnology is required.

Regardless of the debates and controversies, despite its potential benefits and harms, nanotechnology is here to stay and enjoy an exponential growth – for this scientific field promises many inventions on the near and distant horizons hopefully to make mankind's life easier on this planet Earth.

CHAPTER FIVE

Strategic Management Trends in Cyber Age Economy

Fᴏʀ ᴍᴏsᴛ ᴏʀɢᴀɴɪᴢᴀᴛɪᴏɴs the transition from the twentieth century to the twenty-first has had little significance in terms of change of operations. However, the new millennium has ushered in new conditions that necessitated new strategic planning and management philosophy. The massive managerial advances (known as the re-engineering or re-structuring era) brought about in the 1990s were quickly overshadowed by first the collapse of the dot com economy and secondly by the September 11, 2001 destruction of the World Trade Center in New York.

Loss of Faith in the New Economy

Both of the foregoing events resulted in the loss of faith in what was called the "New Economy". These events have also created diffidence in institutions and economic agreements that had dominated international relations for over half a century. As a result, the fall-out for the business world was considerable. The world economy was plagued by a recession in the early 2000s and the U.S. Government's war on terrorism resulted in increasing trends towards "unilateralism" which consequently reversed many of the trends towards closer integration of the world economy.

All of these events, whether domestic or international, imply that planning for the future is becoming harder and harder. Strategic planners across the spectrum need to re-assess their traditional planning processes. The need for new means or tools of scanning the different systems operating in any organization's external environment are becoming increasingly important.

Due to space constraint, only a few recent developments in strategic planning processes and their implications for both strategy scholars and practitioners are observed here. These implications are hoped to serve as forums to strategic planners in generating new ideas that more than likely will reshape our thinking and strategies on how to cope with this era of high uncertainty and rapid change.

Recent Trends in Strategic Planning Processes

Trends in Digital Economy. Alan Greenspan dubbed the phrase "irrational exuberance" in the late 1990s in describing the new economy of the dot com era. The shift from industrial economy to Cyber Age economy where intellectual capital was replacing brick and mortar assets had created a new challenge for strategic planning. The new economy of the late 1990s has been described as the "third industrial revolution". Different companies rushed to capitalize on this new asset of knowledge or intellectual capital. Traditional strategies of lowering cost or taking advantage of economies of scale were no longer the first choice for managers. On-line methods (business-to-consumer or business-to-business) have challenged traditional marketing and distribution strategies.

As Alan Greenspan put it in a speech: "In today's world, where ideas are increasingly displacing the physical in the production of economic value, competition for reputation becomes a significant driving force, propelling our economy forward". Thus, the digital economy will make it imperative for companies to lean heavily on their good reputation to be successful in this hypercompetitive marketplace.

Trends in Competition. One of the consequences of the digital economy and its information technology is the ever increasing competition which has been dubbed as "Hypercompetition". The electronic commerce and the Internet of the digital economy have created a greater competition than existed during the past economy. A hypercompetitive economy has to depend heavily on the generation of new resources for the firm, such as new products, services, procedures, and processes.

The Internet abolished some of the traditional barriers to entry in a number of industries. It also resulted in creating instant global markets for little known businesses. It has created a new, for less costly, distribution channel which drove cost of distribution to minimal levels compared to traditional distribution channels. The creation of electronic commerce has profoundly changed rules of the game for competitors. Weak entry and exit barriers, instant global and national markets, and increasingly price transparency intensified competition. This Internet era has made price competitive variables in many industries.

Trends in Value Orientation. The main quest of any strategic planning system is to match organizational resources and objectives with opportunities found in its macro environment. A major force in the external environment is the value/social system of the immediate and greater community(ies) where that organization conducts its business transactions. Compatibility between interests of business organizations and interests of their community or society should be the goal of any socially responsible business.

The emphasis of the traditional managers lied on the maximization of shareholders' value needs to be replaced by the maximization or fulfillment of society's interests. The recent collapse of some high profile U.S. firms, such as WorldCom, Enron, Global Logistics, and Health South are primary examples of the failure of these companies in reacting positively and quickly to changing societal values and beliefs. Evolving societal values in this digital economy are rapidly changing. This trend has created a challenge and a need to strategic planners to be able to predict such changing needs and values in a timely fashion.

Trends in Manufacturing Strategies.

Manufacturing in its traditional mode where company's assets are mostly invested in heavy equipment and other factors of production is on the decline. More than half of capital investment in the U.S. is in information technology. As a consequence, innovation occurs at a rapid pace. Not surprisingly, the U.S. manufacturing jobs have been in a sharp descending trend since the early 1990s. The digital economy has eliminated manufacturing all together in some industries. A new form of companies called the "virtual corporation or organization" is on the rise. To enhance flexibility, companies have to rely on outside independent firms. Almost 95 percent of virtual company operations or processes are outsourced.

"Trends in Strategic Planning." In the pre-Internet economy, the premises of strategic planning were somewhat simple: low inflation, reasonable consumer demand, stable regulations, low interest rates, and stable equity markets. This stability in the business environment enabled strategic planners then to proceed with their aggressive strategies such as re-structuring, downsizing, outsourcing, refocusing, and engaging in drastic cost cuts. Those strategies resulted in substantially improving the bottom-line results of many large firms.

As we entered the twenty-first century most of the aforementioned strategies had already been exploited to the limit. The substantial gains in profits of the late 1990s began to disappear. As a result, firm's valuations or stock prices suffered. Some executives responded to demands and high expectations by their shareholders in unethical approaches, such as, what was known then as "creative accounting." In the short run, they employed creative accounting measures which kept valuation high. At some point, those firms ran out of any new or creative scheme to artificially boost revenues and profits. Well, by now we all know that that the path chosen (i.e., creative accounting) was the wrong strategy to follow.

Serious and responsible firms opted to take a step back and think strategically of what can they do to reverse their fortunes. A new strategy of focusing on what they know best or focusing on areas of their business where they have obvious and clear competitive advantages is gaining popularity. This refocusing strategy is also extended into other activities such as mergers and acquisitions. It seems that there is a trend towards consolidation in many industries. However, this consolidation is different from the one we saw in the 1990s where high regulations enabled some firms to buy businesses about which they know very little.

According to Peter Drucker, the diversity of product and service contributions by a company will make organizations ". . . fashioned differently: for different purposes, different kinds of work, different people, and different cultures." Today's refocusing merger strategy is intended to enlarge or build on exiting core capabilities and not to gamble on acquiring new unknown capabilities. The drastic failures of some Internet era mergers, such as the Time Warner/AOL merger have made management very leery of pursuing mergers unless such mergers meet their very strict acquisition/merger criteria.

"Trends in New Management Operations." The success of the 1990s was largely based on a model of performance-based management aimed at maximizing shareholders' value. The aim of maximizing shareholders' value was facilitated by the wide-spread use of financial incentives such as stock options. These financial incentives have resulted in unimaginable gains in shareholders' value by the end of the 1990s. This form of management incentives has spread to other similar economies around the world such as Germany and Japan. However, this widespread use of management incentives with the quest for shareholder value has created numerous cases of corporate greed and other corrupt practices. These issues have undermined the credibility of the successful management incentives model of the 1990s.

As stated above, today's business environment possibly exhibits greater dynamism and greater complexity. It is unlikely that the management model of the 1990s will fit today's business environment. The issues on which strategic planners need to focus in today's environment are far more challenging. One of the issues which need an effective reconciliation is the trade off between short-run vs. long-run profit optimization. Management has ignored this issue in the 1990s and opted to focus exclusively on maximizing short-term gains. The new model of management of this new millennium calls for taking this issue seriously. Evidence from numerous corporate scandals of the late 1990s showed that short-term gains are short-lived. To maintain a sustainable shareholder value, management needs to re-focus its attention towards achieving long-term goals.

Trends in New Breed of Strategy Managers. The era of restructuring and shareholder value focus has been associated with traditional leaders who were highly visible and major drivers of change such as Bill Gates of Microsoft or Steve Jobs of Apple Corporation. Such leaders have been, first and foremost, agents of change, strategists, and innovators of their companies. The dynamic and complex business environment of the 2000s requires different strategic skills from management than the above mentioned traditional ones.

To improve the quality of the decision-making process, managers are using decision technology, the application of decision-support modeling and computer soft-ware to their problems in business. The whole process is a value added activity to enhance success by lowering uncertainty surrounding different courses of action. Some recent studies on the evolving role of the new leader, have suggested that the new strategy manager needs to shift his skills from masculine to feminine ones. Traditional masculine attributes such as strong decision making, leading the troops, and taking initiatives are quite different from feminine skills such as great listening capacity and the ability to nurture and build relationships.

The emerging realities of strategic planning processes cannot be ignored in the new landscape of world economy. There are many other areas to be considered. For space consideration, we have only scratched the surface of the changing environment of business both at home and abroad. Additional skills are needed in today's business environment to address issues of diversity in our work force. Strategic managers need to focus on building organizations composed of teams rather than individuals. Focusing on teams tends to mitigate the effects of individual and cultural differences. It is possibly more worth-while for strategy managers to focus on building the right teams over creating the right strategy. After all, as they say "Individuals play the game, but teams win championships."

CHAPTER SIX

Sustainability and Social Cost: The Becoming of a Sexy Subject in Business

IMPERCEPTIBLY BUT STEADILY, science and technology have shaped the world in which we all live today. Twenty-first century world would look rather antediluvian without inventions such as harvesters, automobiles, telephones, and computers.

The scientific strides made in the last one hundred years have in most cases irrevocably affected humans in the way we think about ourselves and in the way we treat the environment. We have progressed from a sense of uncertainty about life on Earth to the confident exploration of space. The notion that God has created the world in certain divine ways is being challenged by the scientific notion that we are the undisputed masters of our destiny on Earth.

Social Cost in Aristotle's Time

Today, slowly but surely, we are experiencing the backlash of science and technology. Global warming, degradation of the land, chemical warfare, air pollution, avian flue are but a few of the side effects of the industrial revolution.

We have come face to face with the realization that our progress has a dear social cost to affect us all on the globe.

The study of the economy in Western civilization began largely with the Greeks. The two luminary figures were Aristotle and Xenophon. According to ancient philosophers, social costs were more clearly discernable, especially in usury. The contention was that the absurd mathematics of compound interest increased social inequality, reduced free men into indentured servitude, and burdened civil authorities with enforcement. The only advantage it seemed to bring was to encourage consumption, which was regarded to be a morally undesirable activity. Thus, interest-bearing debt was not only unnatural, but also a morally repugnant and socially detrimental practice. As we can see, the concept of social cost is very old and has been taken into consideration in early economics.

Social Cost in Modern Times

Social cost, in economics today, is the total of all the costs associated with an economic activity as is shown in Figure 1.

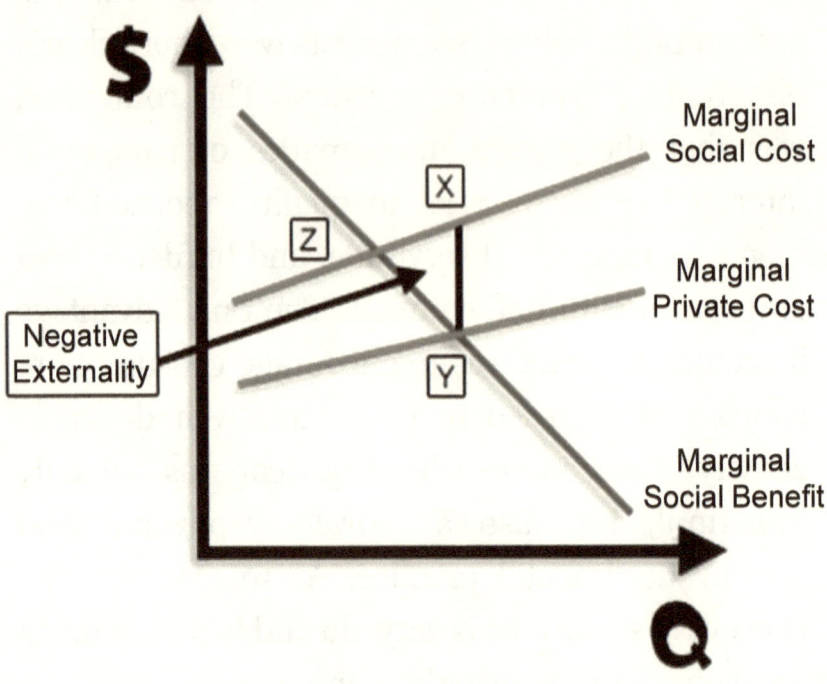

It consists of costs borne by the economic agent such a manufacturer and all costs subsequently borne by society at large. It includes the costs reflected in the organization's production function (i.e., private costs) and the costs external to the firm's private costs called externalities or external costs. In the event, social costs are greater than private costs, then a negative externality emerges. The diesel truck emission pollution is an example of a social cost that is seldom borne by the polluter thereby creating a negative externality to be borne by fellow motorists on the highway (i.e., by society at large).

On the other hand, if private costs are greater than social costs, then a positive externality exists. For instance, when a textbook publisher indirectly benefits society as a whole, and yet only receives payment for the direct benefit received by the recipient (the student) of the education. Thus, the benefit to society of an educated citizenry is a positive externality. In either scenario, economists refer to this as "market failure" because resources would be allocated inefficiently (the marginal rate of transformation in production would not be equal to the marginal rate of substitution in consumption due to the effect of the externality; as a result, optimality would not occur). Even though optimality is elusive in economic theory, generating positive externalities will benefit us all as citizens of the world.

Admittedly, some of the costs of economic growth cannot be measured in terms of national income. Nevertheless, we need to estimate the debilitating effects of such growth which presents hidden problems of congestion, over population, international tension, and social problems in the form of drugs, crime, and homelessness.

For example, the automobile industry in the United States has benefited the economy for over a century in terms of employment and national income. Despite the earnings, the costs of building roads, policing traffic, and treating of diseases associated with air pollution are to be borne by society.

The Burden of Social Cost is Carried by the Consumer

Many social costs are passed on to the consumer in a way that distorts the workings of the marketplace. For example, farmers try to boost crops by using pesticides and chemical fertilizers which eventually drain into rivers and aquifers. As a result, the water is polluted and becomes anything but potable. The farmer does not pay for the cleanup. The task falls squarely on the shoulders of the city or the state to purify the water. Ultimately, the consumer ends up paying for it in the form of increased water bills. Farmers, who grow food organically and do not pollute the waters, do not benefit from such a subsidy. So there is no incentive to produce products friendly to the environment.

There is also another problem in the assessment of social costs. Economists are unable to statistically compute social costs. While it would be easier to assess the contributions of a firm to the national income, it is difficult to assign negative costs to the environment within which it operates. The degradation would include the killing of fish and depriving the community of enjoying the waters surrounding their towns and cities for recreation. The delayed effect and the ripple effect of pumping products into the marketplace make it difficult to anticipate or measure social cost in advance.

Sustainability of the Environment

Sustainability of the environment largely depends on our learning how to quantify social costs and how to make polluters pay for them rather than passing on the costs to the consumers. Against the backdrop of increasing rush to globalization, the world resources are threatened by local as well as foreign companies. Globalization is driven largely by market economics; its environmental social costs are staggering. Unless we factor in social cost into the formula of assessing the benefits of producing private goods, the sustainability of the environment will be jeopardized.

The global community is slowly changing its attitude toward world resources. For example, the beautiful Blue Mosque in Istanbul does not solely belong to the Turks. Similarly, the Taj Mahal in India does not belong to the Indians. These are world treasures to be enjoyed by everyone. Likewise, the environment in Brazil does not belong to the Brazilians alone; the world has a stake in preserving its well-being. For example, the rainforests of the Amazon Basin is essential to everyone on the planet Earth.

The workings of the ecosystems affect us all now – no matter how far we are located from one another. The Earth is like a giant tree: Although the tree thrusts its tap root deep into the back yard of a house, its feeder roots transcend the boundaries and spread into the yards of the surrounding homes. Interrelated and interdependent are all the peoples of the world.

Back to Pigovian Taxation

Perhaps the time has come for the Pigovian taxation. Because the market mechanism fails to factor in the total cost to society, output decisions are flawed, resources are allocated inefficiently, and social welfare is reduced. One method of reducing the effect of this market failure is to impose a Pigovian tax equal to the amount of the negative externality (or impose a subsidy in the case of a positive externality).

Although business colleges are not producing dilettantes when it comes to environmental issues, the discussion of social cost is usually cursory in the business curriculum. Naturally, such an important topic deserves more than a mere paragraph or two treatment in our business textbooks. We can no longer afford to let industry proliferate the market with private goods without being also responsible to take care of social costs involved with their products.

The Societal Marketing Concept
to the Rescue

Ideally, though, business and industry should practice the societal marketing concept whereby only offering products that have no or minimum social costs and, in this way, avoid subjecting the environment to negative externalities. Central to the discussion of social cost should be the use of preventive approach rather than concentrating on methods to rectify the damage done to the environment. We all have the elasticity of the mind to see that it would be counterproductive to try to recover from the diminution of the environment. Our dilatory attitude made us lose large territories of the rainforest around the world.

No great imagination is required, either, to see that academicians as well as business executives have to try to come to grips with this elusive quantification of social costs. Somehow we need to make the subject of social cost sexy to our business students as well for they are our future executive decision makers. The need is long overdue. While economics may be dismal, with so many talented educators we have, sustainability and social cost can become a sexy topic in business.

CHAPTER SEVEN

The Profound Legacy of Science and Its Primary Beneficiaries

HOMO SAPIENS' QUEST for answers to "how?" has led the human imagination to the farthest reaches of outer space. The deep passion and the unquenchable thirst for knowledge have led humankind develop the elasticity of the mind to question reality even beyond outer space.

Science: The Shaper of the World

Through a realistic and literal sense, science has imperceptibly laid the infrastructure essential for contemporary life. In other words, it has slowly shaped the world in which we live today. Had it not been for numerous inventions such as the automobile, the telephone, the skyscrapers, the computers and the like modern life would have looked rather primitive on the planet Earth. We would have continued to till our fields with a wooden plow and a pair of oxen.

Science has also had its effects on humans in the way they think about themselves. In the course of almost 300 years, humankind has begun to appreciate the vastness of the universe on one extreme end of the continuum and the finite minuteness of subatomic particles on the other end. Both of these kinds of scientific explorations have produced permanent effect on our understanding of our place in the complex scheme of things.

Science has altered some of our long cherished medieval notions of ontology in that humanity had been made in the image of God at the center of an ordered universe to the realization that humans are all made up of the same matter as the stars in the sky and the Earth on which we live. Moreover, science has provided us with our current state of consciousness through a series of random mutations. Despite humankind's immense scientific progress, we have only scratched the surface of our elusive and mysterious world.

Change of Attitude Toward Love

Such a fundamental change in attitude required many years of speculation, research, and discoveries. For example of great importance in humankind's life is love. With the exception of courtly love, up to medieval times, love was considered to be an affliction not to speak about it in public. Though love is complex in any age, romantic love (passionate relationship between a man and woman) during the medieval period consisted of two major elements that are entirely at odds with each other: suffering and pleasure (i.e., pain and lust!).

Today, psychology has recognized that "love" is essential for the wellbeing of the individual and for the survival of the species. Women have to feel fulfilled by bearing children and men by fathering them. Although psychologists in defining love differ, most agree that it consists of three major components: Passion, Intimacy, and Commitment. The presence of these ingredients between a man and a woman, render the relationship as a solid, enduring, and a "Consummate Love," (the best of its kind). Studies have shown that all of these three elements foster bonding and a sense of belonging. Hence, in modern times, love is hailed as "a many-splendored thing" as the popular song states and not as a "dreary thing" to keep it secret as a personal curse as believed in medieval times. Science is changing our perspectives on many issues as well too numerous to cite them here.

Natural Philosophy as the Primary Science

Before the Scientific Revolution of the 17th century, Natural Philosophy, as the primary science of the day, consisted of a number of "certainties" handed down from the ancient

Greeks. For instance, the heavens were made up of seven crystal spheres. In the age of Renaissance, new inventions, such as the telescope and the microscope, began to expose the hidden complexities of the universe by empiricists such as Newton's experiments showed that the world was not always quite what it seemed to be.

By the end of the 19th century, the general sentiment was that by using the magic of science we could know everything and so strengthen our position as "masters of God's universe." A whole new age of industry had been built on scientific discovery: the combustion engine, the railroad, electricity, the telegraph, etc. Over the years, business and industry, thus have been the primary beneficiaries of science. Yet seldom, if ever, business professionals and university business professors engage in writing about science, the major source of their products and services. They should, in fact, engage in research and writing on the many exciting and wonderful facets of science. After all, business and industry are the two major and undisputed beneficiaries of science.

The Aftermath of the Atomic Bomb

Unfortunately, the misbegotten atom bombs dropped on Hiroshima and Nagasaki in 1945 delineated the turning point in the optimism of science's noble contributions to human progress. Ruefully, we have come to realize that we had become the masters of our planet – the masters of destruction as well by using the tiniest particles known to humankind we had created the fastest catastrophe the world has ever suffered.

Like most things in life, not everything is benign. The backlash against science is mounting. Science has been blamed for its rapid dive into the future; it has also been condemned for ushering in the proverbial calamities of pestilence, famine, and war. To many observers, global warming, chemical warfare, the threat of nuclear fallout, new diseases such as aids, the Ebola virus, are the results of scientific endeavors to push the envelope by going one step farther. Despite the dark side of science, it is through science and the application of human brain that these problems may someday be resolved.

CHAPTER EIGHT

The Impact of the World Wide Web on Business and Society

WHEN IT COMES to exploring history's greatest breakthroughs and inventions which have had a profound impact on business and society, one idea is often ostensibly absent from the list. This often overlooked idea is the invention of the World Wide Web (WWW).

A Brief History of the World Wide Web

The brief history of the WWW goes back to 1980. It began in the Swiss Alps by Tim Berners-Lee, a British software engineer working temporarily at CERN, the European Laboratory for Particle Physics, in Geneva. One day, as Berners-Lee was fooling with a way to organize his far-flung notes. Based on the prevailing ideas in software design at the time, he designed a kind of "hypertext" notebook. Basically, words in document could be linked to other files on Berners-Lee's computer.

To open up his document and his computer to everyone else and allow them to link their stuff to his, he cobbled together a coding system called HTML (Hypertext Markup Language) and designed an addressing scheme that gave each Web page a unique location, or URL (Universal Resource Locator). By hacking a set of rules that permitted these documents to be linked together on computers across the Internet – HTTP (HyperText Transfer Protocol), WWW was born.

Within a few weeks, Berners-Lee assembled the World Wide Web's first browser which allowed users anywhere to view his creation on their computer screens. Later on, Berners-Lee alerted the computer users through a message posted to a newsgroup about the new service and the world embraced his invention akin to love at first sight.

The Rise of the WWW

On August 6, 1991, the Web officially made its debut, thus bringing instantly order to the chaos that was cyberspace system of scattered, but related, documents. From that fateful moment, the Web and the Internet married as a couple and grew as "husband and wife" at geometrical rates.

Within five years of its inception, the number of Internet users jumped from 600,000 to 40 million in 1996. Until then, we had no idea what a powerful new tool the computer could be for everyone through the creative collaboration and melding of the Web with the Internet. Ever since then, the two have had a happy and prosperous marriage because the Web introduced sex appeal through colorful documents into the dull life of the Internet!

The Web officially launched as an offshoot of the Internet in 1989, has become a huge part of many people's lives. The invention enabled them to communicate, work, and play in a global context. Essentially, the Web is all about relationships, and has made these relationships possible between individuals, groups, and communities where they would not have been possible otherwise. It has been nothing but love-love relationships between the Web and its loyal users. One of the major impacts of the Web has been in creating a community without borders, limits, or even rules; it has become a true empire of its own.

The Major Contributions of the WWW

Very briefly, the major impact or contribution of the WWW has been in the area of effective communication. More than any technical definition, the Web is a way that people communicate.

More than anything else, the Internet made people realizes that communicating by snail mail was less effective and more costly than free e-mail on the Web. The possibilities of Web communication was mind-boggling to people when the Web was just getting started.

Nowadays, we think nothing of e-mailing our friends in France and getting an answer back within minutes or simply seeing the latest video full of the latest news. The Internet and the Web have revolutionized the way we communicate, not only with individuals, but with the world as well.

Life Without the WWW Would Be Unimaginable

It would be hard to imagine life without using the Web: no e-mails, no access to breaking news, no up to the minute weather reports, no way to search for products, no way to find information about gifts, no way to shop online, etc.

The Web has been now hardwired in our genes; we have grown to be dependent on this technology; it has transformed the way that we conduct our lives. Without the Web, we would act as though suffering from bipolar disorders, Schizophrenia, or from manic depressive tendencies.

The Web has become a movable feast for most of us. Like the color of the chameleon, the Web cannot actually be tracked down. It is a continual, ongoing process. It never has stopped replicating itself or progressing since the day it was conceived and born. Most probably, it will keep evolving as long as people like Berners-Lee are around to keep developing it. In a nutshell, the Web is made up of personal relationships, business partnerships, and global associations. Without these interpersonal relationships, the Web would cease existing.

CHAPTER NINE

Toward Taming the Monster in Electronic Mail

WHENEVER A NEW technology emerges, the tendency has been to embrace it wholesale without considering seriously the implications for its use in the future. What might seem to be a blessing today may very well become a curse of tomorrow unless drastic measures are taken to tame an innovation such as electronic communications which has morphed into very objectionable formats.

A case in point is the use of e-mail for corresponding with students, colleagues, customers, voters, and the public and the strife to survive the onslaught of "spammers". While, in the beginning, a few e-mails from any source were a welcome sight for using the new technology, a recent survey had disclosed that most professors, for example, are getting an enormous amount of e-mails to the tune of 20 to 30 e-mails a day from students and other sources.

The Pandemic Problem of Electronic Mail

The problem has become pandemic in the academic, business, and government affairs. Under such an avalanche of correspondence, the average college professor succumbs to the pressures of responding to e-mails and in this way relegates his or her research and writing. Most of the unwitting Internet users of the world are experiencing e-mail explosion. To explain the magnitude of this explosion, the following fable puts the problem into perspective:

Once there were two kings from Babylon who enjoyed playing chess, with the winner claiming a prize from the loser. After one match, the winning king asked the loser to pay him by placing one grain of wheat on the first square of the chessboard, two on the second, four on the third, and so on. The number of grains was to double each time until all 64 squares were filled. The losing king, thinking he was getting off easy, agreed with delight. As you may know, it was the biggest mistake he had ever made. He bankrupted his kingdom and still could not produce the 2^{63} grains of wheat he had promised. In fact, it is probably more than all the wheat that has ever been harvested!

This parable is an example of exponential growth, in which a quantity increases by a fixed percentage of the whole in a given time. As the losing king learned, exponential growth is deceptive. It starts off slowly as e-mail did in the beginning, but after only a few doublings it grows to enormous numbers because each doubling is more than the total of all earlier growth.

Despite the strain that spam and e-mail alerts has been putting on "in-boxes," market researchers of IDC has predicted even more pressure from e-mails exploding from ever-increasing rate in the coming years just like the exponential growth episode of the above fable: The overall number of e-mail messages will double from 31 billion a day in 2002 to a staggering 60 billion a day by 2016. The enormity of the problem is readily discernible despite the fact that what we have been so far experiencing is the tip of the iceberg.

Two World Revolutions According to Gates

According to Bill Gates, the world has recently witnessed two major revolutions. The first one is the PC revolution; the second one is the Internet revolution. The former revolution has been a panacea for fast computing, while the latter revolution has somewhat become problematic, especially in the area of electronic mail as he called the phenomenon "The Internet Tidal Wave." And tidal waves are destructive, we all know, and show no mercy to those who happens to be in their way. The euphemism here is apt for the unsolicited, mass-transmitted e-mails sent out to many recipients.

Considering the e-mail as one of the most important attributes of the Internet, Bill Gates conjectured further that corporate structures would evolve because e-mail was a powerful force for flattening the hierarchies germane to large companies. When communications systems become effective, companies would not need so many levels of management. Intermediaries in middle management, who once passed information up and down the chain of command, would no longer be as important today as they once were.

In a sense, because of e-mail, there are currently no levels between Bill Gates and anyone else in the company. Despite Gates' euphoria over the attributes of the Internet, little did he expect that the e-mail explosion would become a pandemic problem for the academic, business, and government sectors.

The Digital Addiction

With the explosion of the digital workplace, e-mail has become an integral part of the infrastructure for the way in which we communicate nowadays with colleagues, business partners, customers, suppliers, and friends and family members. While this medium for communication has presented many benefits with which we have become familiar, such as with its speed, informality, and postage-free mail status there are some inherent liability concerns that could be directly attributed to the use of e-mail such as overwork and stress.

Recently a researcher experimented for 20 days without a PC. He found out that his entire world had changed; it was difficult for him to survive. On the other hand, there has been a rising chorus of complaints against unsolicited e-mails by both students and professors. Their vociferous complaints have failed to convince the college administrators to find ways to fend off specially spam e-mails.

Unlike the U.S., for instance, Australians seem to be holding the bull by the horns. The Independent Education Union of Australia has been pursuing a pay increase of at least 20 percent for teachers and staff for time spent answering students' e-mails. The difference now with e-mails is the students' expectation that the teacher is on call twenty-four hours, seven-days a week (24/7). As a result, there is a massive increase in workload which is eating into private time for leisure or for research and writing.

Most businesses and households have evolved from 'snail-mail' and have been now using e-mail as their primary form of correspondence. This explosion in e-communication has brought tremendous benefits for marketers, but, unfortunately, it has also ushered in serious risks and hazards involving security, hackers, virus, and computer down times.

The Dark Side of Electronic Mail

Some lament that there has been a "darker side" to the world of e-mail. Civility seemed to have been vanishing in our communications. Because it has been relatively so fast and easy to use, people have become deluged with e-mails. As a result, everyone has become overworked and stressed out. Only few people have the time to write long, frequent, friendly, helpful, or even clear letters.

Over the past few years, much attention has been placed on the consideration of finding ways to help the public cope with the onslaught of e-mail correspondence, especially for dealing with spam e-mails. We have all been waiting for the technology sector to tell us, in no uncertain terms, how will we be able to control the free flow of e-mails. Incidentally, on March 21, 2005, IBM announced plans to launch a new anti-spam product which would return "junky" mail to its sender.

Given the criticality and pervasiveness of the problem throughout our society, each and every individual should make an effort to create awareness of the current methods available to curb the epidemicity of e-mails in our daily life.

Legislation to Curb Unwanted E-Mails

In an attempt to stem the flood of unwanted e-mails, President Bush signed legislation in 2003 against unwanted e-mail pitches, which are believed to be annoying and costly to our economy.

The White House spokesman Scott McClellan said that "This [the legislation] will help address the problems associated with the rapid growth and abuse of spam by establishing a framework of technological, administrative, civil and criminal tools and by providing consumers with options to reduce the volume of unwanted e-mail."

Supposedly, spammers will now face tough rules and harsh penalties for sending unwanted, offensive e-mails to unwilling individuals. The bill supplants anti-spam laws already passed in California and in some other states. The legislation also charges the Federal Trade Commission to create a do-not spam list of e-mail addresses and includes penalties for spammers of up to five years in prison in some instances. So far the results of "can spam" legislation, which outlaws the persistent techniques used by e-mailers who send tens of millions of messages each day to peddle their products and services, have not lived up to the expectations of the bill.

While spasm has been the main pestering problem, e-mails from other sources have also contributed their lion's share toward the explosion. Students have formed the habit of "shooting" e-mails to their professors anytime of the day or night. They should learn to be considerate and responsible users of e-mails. For example, instead of students sending e-mails inquiring as to when the midterm exam is scheduled, they could easily tap into information readily available at their finger tips, namely consult the course syllabus. In so doing, the teacher would spend less time and effort on responding to e-mails from students every day and spend more time on curriculum and research matters.

The Need for Educating E-Mail Users

There is a great need to educate our students, the business sector, the government, and the public to refrain from adding to the explosion with unnecessary e-mails. Among the biggest offenders are the spammers who are polluting our digital environment day in day out. While legitimate e-mails are essential now for fast interface, redundant and spam e-mails are counterproductive for the society and the economy as a whole.

It has become utterly imperative to exercise parsimony and to avoid redundancy by searching easily available information sources first before automatically resorting to e-mail communications. Taming the monster in the e-mail correspondence should be the responsibility of every Internet user. Of all the technology-based approaches, nothing would be complete unless the Internet users make a conscious effort to curb the explosion. The key is exercising self-restraint from generating frivolous and unnecessary e-mails.

Proper awareness and education on the "darker side" of e-mails can be achieved through journal articles; special round table discussions; IT and Marketing and Management tracks at conferences; seminars; symposia; advocacy ad campaigns, etc. Studies, articles, and reports on how to control the monster in the e-mail correspondence would be highly appreciated by the public.

We all need to be staunch supporters of the activists who address the issues of ever-increasing, vexing, and time-consuming onslaught of the spammers' relentless offers. May the wind be always behind our activists and may our "In-box" be – unlike the losing Babylonian king's chessboard squares – free of any pesky, multiplying "grains" of spam.

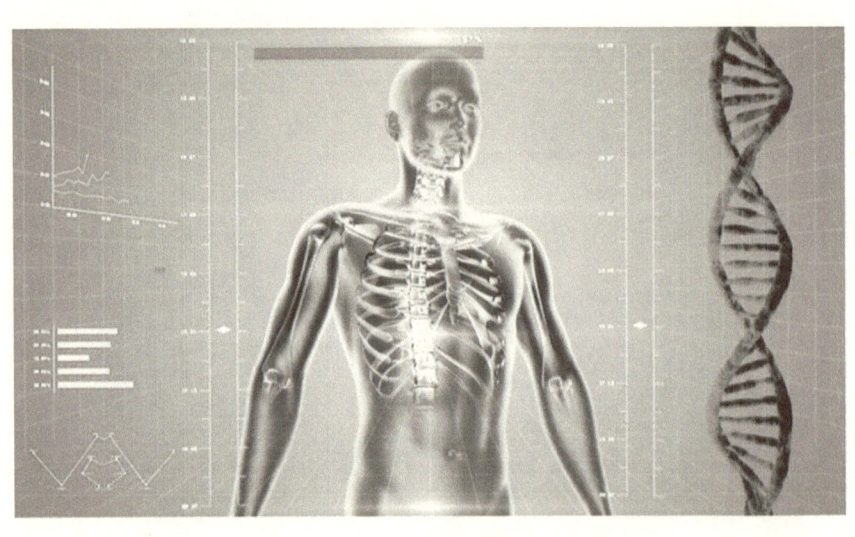

CHAPTER TEN

The Genomic Revolution: Gifting a Generous Harvest

THE GENEROUS HARVEST yielded from the genes research and later from the genomes technology is having a growing and an important impact on our lives. Whereas genes are the basic unit of heredity in all living beings on Earth, genomes are the entire set of an organism's genetic material. Genomics has emerged from the melding of many disparate fields, such as biology, public health, engineering, computer science, and mathematics.

Social sciences, such as Marketing, Management, IS, and humanities are also an integral component of the genomic revolution.

The Long History of the Study of Genes

The study of genes dates back to well over a hundred years. Not long ago, as great strides were made possible by sequencing of the complete genetic structure of an organism (i.e., genomes), the quest shifted to genomic discoveries which has captured the attention of the world scientists and as well as world entrepreneurs ready to cash in on new products and services.

Milestone research accomplishments span from the 1950s to the present after having identified the key to heredity. In 1953 DNA structure was established by James Watson and Francis Crick. Within a dozen of years, the genetic code was cracked by Marshall Nirenberg and his associates in 1966. In 1972 the first recombinant DNA molecules were produced by Paul Berg and in 1973 Stanley Cohen, Annie

Chang, and Herbert Boyer produced the first microorganism with recombinant DNA.

At the University of Ohio, the first report was issued in 1982 regarding the successful production of transgenic animals, mice, etc. In less than a year, in 1983 Luis Herrera-Estrella and associates reported of genetically modified plant cells. In 2001 Shatten and associates shook the scientific world by producing the first genetically modified primate, a rhesus monkey called ANDi from the reverse initials of "inserted DNA". All of these developments were based on technologies in recombinant DNA, transformation, cell and tissue culture, gene transfer to mention a few.

The aforementioned technologies gave rise to genetic engineering to produce genetically modified organisms (GMO) in plants, animals, or microorganisms and paved the way to the commercialization of many products. Thus, the move from genes to genomes has introduced whole new technologies (post-genomics) which are geared for genome-wide analysis of gene structure and expression, including computer-based analyses of vast data sets known as bioinformatics.

In the last thirty years, a revolution has occurred that has catapulted molecular biology at the center of all the biological sciences, and has had wide-ranging implications in many fields, including business and political landscapes. The major force behind this revolution was the development of methods that allowed the isolation of specific DNA fragments and their replication in bacterial cells known as gene cloning. These methods enabled researchers to engineer bacteria, plants, and animals to have novel properties, including the production of pharmaceutical products.

Contributions of Gene and Genome Research

Contributions of gene and genome research constitute a generous harvest. Years of research and development have produced a plethora of benefits. There are two major categories of potential benefits: medical and non-medical.

The potential for medical benefits include the production of human therapeutic proteins; progress in the study of genetic diseases using GM animals as models; progress in modifying animal organs for human transplants (xenotransplantation), and production of offsprings completely different from either parent (xenogenesis).

The potential for non-medical benefits are more numerous: recombinant enzymes in food use, higher yields, disease resistant, and improved nutritional and processing quality in crop plants are a few examples in enhanced agricultural production. Evidence based on DNA testing are more and more frequently used in criminal cases. To a lesser degree, scientist are applying DNA data to determine a person's or a society's relationships to other people through genealogy. Finally researchers are using DNA readings to determine the origins of an industry, such as the cradle of winemaking in the ancient world and traceability of products.

Applications of Genome Technology

Applications of genome technology can be grouped in four major areas of potential benefits in medicine and non-medicine, respectively.

Benefits in Medicine:

❖ Enhancing Health Care – Human therapeutic proteins including basic material such as bacteria and yeast can be genetically engineered to produce mass quantities of biological products such as human insulin, growth hormone and hepatitis B vaccine.

❖ Studying Genetic Diseases – Genetic screening, they say, could reveal vitally important information about a person's life span and health prospects. Such screening already identifies certain diseases that run in families, enabling couples to decide whether or not to have children.

❖ Applying Xenotransplantation – Modification of animal organs for human transplants involves production of biological products. This whole thing is also dubbed as "Biological factories". The Cambridge-based Imutran is one of the companies now breeding pigs with a human gene in an attempt to create animal organs that will not be rejected so easily during human transplant operations.

❖ Conducting Exogenesis – It involves the production of unique offsprings by gene grafting. The first animal to be patented, in the U.S. during 1988, was Harvard University's 'oncomouse', designed to develop cancer. The patent applied not just to mice but to any non-human mammal with an inserted oncogene. Although several other transgenic animals have since been patented in the U.S., the situation in Europe is yet to be finalized.

Benefits in Non-Medicine:

❖ Detecting Criminals – DNA analysis has emerged as an extremely valuable tool for the American criminal justice system. In countless cases it has served to identify the suspect, convict the guilty, and bring some solace to the victim. In other cases, it has exonerated the innocent, at times after years of unjust imprisonment. DNA's capacity to illuminate the empirical truth provides the opportunity to use it for enhancing the efficiency, effectiveness, equity, and credibility of criminal justice throughout the nation.

❖ Determining Genealogy – Genes provide windows into the origins of society or specific people such as the relationship of modern day Lebanese with their Phoenician predecessors who dominated sea trade in the Mediterranean for 3,000 years and gave the world the Alphabet, and then mysteriously disappeared.

Pierre Zalloua and Spencer Wells started work three years ago on a study to pursue the reading of the genetic makeup of the Lebanese and populations from the Mediterranean basin where they established colonies. The study concluded that the Lebanese belong to the "older" Mediterranean substratum. This means that the coastal Lebanese people share the same genetic identifiers like the Macedonians, Iberians, Basques, North Africans, Italians, French, Cretans, Jews, Anatolians (the aborigine Turks), Iranians, and Armenians.

Finding Origins of Industries – An expert on ancient wine, Patrick McGovern is searching for the origins of the first domesticated grapevine through genetic technology. As a biomolecular archaeologist, combining archaeology with chemical and molecular analysis, McGovern has already pushed our knowledge of vinicultural history back to Neolithic times.

McGovern was on an expedition to Turkey's Taurus Mountains near the headwaters of the Tigris River, combing the rugged river valleys in search of wild grapevines untouched by modern cultivation. McGovern was joined by José Vouillamoz, from Italy's Istituto Agrario di San Michele all'Adige in Trento, and Ali Ergül, from Turkey's Ankara University. The reason they have been looking in eastern Turkey is because that is where other plants were domesticated. They wanted to collect wild grapevines with local cultivars, so they can see what the relationship is and maybe make a case that this is where the first domestication occurred as stated by many other scholars.

Maple Leaf Foods launched DNA pork traceability program in Toronto, Ontario in 2004. Having successfully completed research and development into producing a DNA traceability program for pork, it launched the first commercial application of this technology in 2004, with far-reaching benefits for food safety and the Canadian pork industry. The DNA traceability system will allow Canadian pork marketed anywhere in the world to be traced back to the maternal sow, providing the Canadian pork industry with a major competitive advantage and an essential point of difference for the "Made in Canada" brand. Canada has set the international gold standard for pork. IBM Canada has the expertise to develop DNA traceability for many key industries around the world.

❖ Improving Agriculture – Techniques have been developed to alter animals' genetic make-up, producing new strains of species to be used by the agricultural, pharmaceutical and biomedical industries. One approach is to insert genes from one species into the embryo of another, the resulting creature is known as 'transgenics'. Another method is to knock out one of the animal's own genes. These creatures are called 'knock-outs'. Much genetic engineering is aimed at getting farm animals to grow bigger and more rapidly.

By virtue of genomics, many crops can be developed that are disease-resistant and plants that grow fast to meet the ever growing world demand for lumber and paper. Indirectly though, journal printing most likely soon become a beneficiary of the genomic revolution. Slick paper is the high definition counterpart of HD T.V. It is not only attractive to look at, it also makes reading easy. It is durable and hence it has longer shelf life than ordinary paper.

Thomas Malthus's Prognostication of Economic Disaster

Many people were alarmed at Thomas Malthus' prognostication in 1798 that there will be worldwide tragedy because population will increase at a faster pace than technology could keep up with it. As a consequence, for humans the problems of disease, suffering, starvation would loom. In the same vein, Mathew Simmons' 2005 book *Twilight in the Desert: The Coming Saudi Oil Shock and the World Economy*; Richard Heinberg's 2004 book *Powerdown: Options and Actions for a Post-Carbon World*; and especially James

Kunstler's pessimistic scenarios in his 2006 book *The Long Emergency: Surviving the End of the Oil Age, Climate Change, and other Converging Catastrophes of the Twenty-first Century* argue that there may not be enough time left to make an orderly transition out of our oil-based economy and thus the future not only looks miserably grim, but it is also cataclysmic.

After 208 years of Malthus' forecast, people of all walks of life still live, some well, some not so well, and some others, unfortunately, starve. Necessity is the mother of invention and science and technology will always come to the rescue – sometimes on time, most of the time a bit late, but always they bail out humanity when faced with difficulties. Genomics seems promising enough to feed and to provide the world with viable medicine.

Genetic Research: Hopes and Fears of Observers

Genetic research is both the hopes and fears of many observers of the science. While the potential benefits of genome technology are admirable, the innovations put in the wrong hands may cause immeasurable destruction in the world. A case in point is the production of organism to be used in warfare or by terrorists.

The metamorphoses of our world are well expressed by Heraclitus (540?-475) long ago by reminding us that "Everything flows and nothing abides, everything gives way and nothing stays fixed" (i.e., one cannot step twice into the same river!). Genome technology will soon launch so many products that the plastic age, the fossil fuel age, would dwarf in comparison. What we already have will soon become Neolithic as genomics advances and as the pressures of increasing population mounts. It has already begun to make inroads into the produce industry. We are on the threshold of the "artificial and/ or virtual age". Most of our plants and animals will be changed through the genome technology to feed the world hunger and to heal the world sick due to the impending world population explosion.

CHAPTER ELEVEN

Solar Energy: The Power to Stop Climate Change

WE AS HUMANS all share one small home while individually we live in separate houses. Our collective home is the planet Earth. Of all the animals, humans dominate our planet. Hardly any place is left where human impact is not felt. Of much concern has been the atmosphere which has been altered by gases produced by humans such as chlorofluorocarbons, resulting from chemical processes alien to natural systems.

Many who commute to work or sail the seas are aware of human debris in our waters. Lakes are drying up, chocked up by pollution as are rivers. The preponderance of world's forests and grasslands has been compromised. Over the last three hundred years, humans have transformed this planet in many ways. In some instances, we have improved life; in others, we have jeopardized existence of societies on this fragile planet. Automobiles have been the arch polluters of the environment and, therefore, are the main cause of climate change with serious repercussions and consequences.

To reverse the damage would be like catching lightening in a bottle. A creative entrepreneur can benefit from the demise of the planet by offering products and services to alleviate the impending effects of climate change. One such product would be solar thermal energy.

The Long History of Solar Energy

Harnessing sunlight into electricity seems to be reinventing itself in California. The idea dates back to ancient Greeks. The concept has been known for 2500 years. In the 1500s, Leonardo da Vinci refined the idea for boiling water for a dye factory with the heat generated by a curved mirror four miles in diameter. Such an impractical mirror was never built, but his mentor, Andrea Verocchio, successfully tested the idea a century ago by devising a smaller one to be used in soldering; therefore, the solar thermal energy fails to qualify as a twentieth century innovation. Since the dawn of time, the sun has been a vital source for life. Therefore, humankind has entertained long ago the idea of solar energy.

Solar energy is derived from the sun in a form of ultraviolet rays. It was first applied by the Greek genius Archimedes in 212 B.C. He used solar energy to defend the harbor of Syracuse against the Roman fleet. Archimedes devised a mirror or "burning mirror" as they had called it to set fire to the ships of the Roman fleets while standing on shore. Since then, it took sixteen centuries for Salomon de Caux to construct the first solar device, a solar engine in 1615. De Caux's contraption was made up of glass lenses, supporting frame, and an airtight metal vessel containing water and air. This produced a small water fountain when the air heated up during operation. This was considered to be more of a toy than a device, but it was the first published account of the use of solar energy since the fall of the Roman Empire. Since then, other uses of solar energy were the solar roof and the solar oven.

Interestingly enough, more than 250 years ago, Comte de Buffon, a French experimentalist, directed 124 flat mirrors on a model ship in Paris and burned it, demonstrating that Archimedes indeed could have burnt the Roman fleet as ancient historians claimed.

In the 1800s, people became worried over steam engines' insatiable appetite for fuel produced by coal of which Europe was on the verge of complete depletion. The industry had to turn to other sources of energy. One scientist went out to build a dish-shaped mirror that focused sunlight onto a boiler to help resolve what he believed was the impending demise of fossil fuels. During the First World War, the first economical solar-powered engine was operating in Egypt which had little in the way of accessible fuel. A few years later, the region surrounding Egypt discovered plenty of oil. Likewise oil was found in North America. The worry about energy subsided from then on, and the interest in solar power dissipated.

Renewed Interest in Solar Energy

Given the high price of oil, the current energy crisis, and the knowledge of the harm burning fossil fuels can cause such as climate change, there is a renewed interest by some scientists, government officials and entrepreneurs to focus on this old, but evolving solar technology as an alternative source of clean and renewable energy.

Advocates for the technology of solar thermal energy have estimated that the world would need less than one percent of all the world's deserts to power the entire globe. This would be a herculean task to achieve, but the successful operation of the thermal plant built 22 years ago in California's Mojave Desert provides some hope.

Early in 2008, San Francisco-based Pacific Gas and Electric company contracted with

BrightSource Energy, Inc. to supply power it plans to generate at three plants in the sunny California desert. Southern California Edison also signed agreements with eSolar, Inc. to buy power from solar thermal facilities to be built in the Antelope Valley (north of Los Angeles). Both of these suppliers expect their first plants to be operational in 2011. Hopefully, this will be the beginning of the end of the age of fossil fuel energy. As they say, man has graduated from the Stone Age not because he ran out of stones, but because he found new technologies such as in bronze, and later in iron.

Why Is the Exploitation of Solar Energy is Slow

Against the backdrop of spiraling price of oil, the burning question is that why then have we not exploited the use of the sun when it gives us a clean and comparatively less expensive source of energy?

There seems to be two main culprits: one is the consumer, and the other is the abundance of cheap oil by cartel's manipulation of the supply of oil and its price. Consumers want cars with muscles and yet economical. They have attitudinal and behavioral problems: one cannot have his cake and eat it too. Cars with muscles must run on fossil fuel which is causing climate change with attendant complications for all the humans on this planet.

Whenever there is abundance of oil such as after the Second World War, the interest in alternative energy sources such as solar power fades away. The impetus behind the search for cleaner, renewable source of energy comes from energy crisis when prices of oil skyrocket. Nations and corporations are not practicing management by objectives, but rather resort to management by crisis. There is a need of attitude change in pursing technologies. New technologies should be pursued also with great vigor during non-crisis time.

What would counter climate change and reverse its dangerous trend? If humankind wants to stop the harmful effects of climate change such as getting one third of Bangladesh (one of the most populous nations) submerge under the sea, humankind has to change first its attitudes and behavior toward the use of fossil fuel. Then as now, solar technology has not been seriously pursued by large corporations. Although solar energy has had a long and rich legacy, no nation has capitalized on it yet. Solar energy is abundant and renewable and requires no more than one percent of world's desert to meet planet Earth's demand for clean and affordable energy. Many firmly concerned citizens believe in the use of energy, such as solar power, which is compatible with the fragile nature of our small and susceptible planet Earth.

Part II

Basic Challenges in the Business World

CHAPTER TWELVE

The Maligned Image of Monopoly

EVER SINCE THE fall of the Soviet Union in the 1990s, one reads in the news now and then that the former Soviet Republics including most Eastern European nations have been devitalized by lack of competition in the marketplace. Oligarchs, large ruling families, had taken over the economy by exercising monopoly and thus controlling the price of the basic necessities of life.

For example, one article recently reported that the price of sugar in one of the former Soviet republics would be raised by twenty-five percent due to an increase of this commodity on the international market. The price was going to be raised by an MP (Member of Parliament) whose family had allegedly monopoly on the sale of sugar in his struggling country.

The article sounded like indicting this oligarch family for taking advantage of the poor consumer who lacked choice. The insinuation, innuendoes, and outright charges denigrated the family and condemned the raising of the price of sugar by the monopolist.

The Negative Image of Monopoly

As you are well familiar with the concept, monopoly (which means one seller) has become a common and even colloquial term in the modern era. Invariably, it has been assigned an almost uniformly negative connotation. If one monopolizes the conversation in a social setting, one is considered to be unwelcome and even rude. If a business maintains a monopoly in their respective field, it is excluding other companies which deserve a fair chance to compete and offer the best price to the consumer.

This last sense of monopoly connotes an inherently unsavory and unethical business practice to the detriment of other businesses and the consumer at large. Monopolies are categorically considered to be counterproductive to the consumer and to the economy. The common contention is that they prevent competition, result in higher prices, sell poorer quality goods, and subject the consumer to the arbitrary whims of a single company. They deprive the consumer to have a choice. So in order to maintain the freedom of the market, one must fight against monopolies and potential monopolies. Pigeon holing all monopolies as being negative would be Quixotic.

This type of indiscriminate prejudice against all monopolies is unwarranted, unfair, and even wrong. Not all snakes are poisonous; not all snakes are dangerous. Likewise, not all monopolies are alike; not all monopolies need to be feared and dismantled.

The Many Faces of Monopoly

A monopoly has many faces. There are multiple types of monopolies classified by economic and legal experts under the same term. Of interest to us here are the two major sorts of monopolies which are differentiated on the basis of a most fundamental distinction: Efficiency Monopoly and Coercive Monopoly.

Efficiency Monopoly can be defined broadly as "monopoly by success." Such a monopoly occurs when a single company (e.g., an oligarch) is so extremely successful, so productive and so efficient that it is able to satisfy customer needs, meet customer standards, and ensure customer satisfaction. In this way, the company becomes a monopoly which maintains dominance in its particular area of the market. By virtue of these characteristics, it can provide a better product or service at a better price than any other company or entity.

The key point is that an efficiency monopoly monopolizes by virtue of consumer choice and by no other means. A case in point is Microsoft

Corporation with its Windows operating system. It has virtually no or fewer competitors because it would not be profitable for any given competitor to enter the company's field. Competition is open, though, it is simply not prudent, cost-effective, or potentially profitable for any other company to enter into competition.

Let us bear in mind that an efficiency monopoly is not unrestricted by certain conditions for it to set prices at an unsatisfactory level, or produce inferior product, there would then exist a demand for competition, and make it profitable for a competitor to enter the field. An efficiency monopoly is bound strictly by the law of supply and demand as it functions in a free market.

Coercive Monopoly, on the other hand, is a monopoly of the sort that consumers, policy decision makers, the legislators justifiably fear because it has an exclusive control of a given field of production or service. The market is closed to and exempt from competition, so that those controlling the field are able to set arbitrary production policies and charge arbitrary prices.

This type of behavior is done independent of the market, immune from the law of supply and demand.

The Distinction between Efficiency and Coercive Monopoly

The distinction between an efficiency monopoly and a coercive monopoly requires further discussion. What is at issue in the disparity between these definitions is how each type of monopoly is attained and maintained. As was previously mentioned in its definition, an efficiency monopoly gains its dominant market-share due to the choice of the consumers by engaging in monopolistic competition. It is however, still subject to the law of supply and demand. It does not forbid competition, but merely makes it hard and even untenable for other companies to enter the market by maintaining its efficiency.

A coercive monopoly, on the other hand, is not subject to the law of supply and demand. It is exempt from competition, prohibits any other competitor from entering the market, and by virtue of this prohibition, it is able to exercise arbitrary power over the market.

An Oligarch Monopolist May Be Efficient

In conclusion, before one passes judgment on a company as being a monopolist due to its size, market share, or price setting practice, and prejudicing the public against it, one should examine first as to what type of monopoly it is: efficiency or coercive? The oligarch which dominates the sale of sugar in one of the former Soviet republics may very well be an efficient monopoly. If that were the case, then we should not use the word "monopoly," but rather an "industry or market leader." This has more positive connotation. Therefore, instead of denigrating the oligarch family, we should be praising it for its efficiency.

While efficiency monopoly is a blessing, a coercive one is a curse for the economy and the consumer. After all, nature is full of examples of the flourishing of the fittest, be it in the animal or plant kingdom. For example, a plant in the forest that works hard by spreading its feeder roots wider for nutrients by trusting its tap root deeper for water tends to grow stronger and greener in the forest while not so hardworking plants tend to languish under the canopy of this tree which dominates the soil and the sun to out flourish the rest of its competitors in a given space. Humans appreciate and admire those entities that "flourish by being the fittest!"

CHAPTER THIRTEEN

Building Relationships by Bowing to Change

TO DISCUSS THE building of relationships in today's competitive world, let us take an example of an academic journal which has attained a relatively popular position among similar publications. Let us take the Journal of the Academy of Business, Cambridge (JAABC) as an example since I am quite familiar with its rapid progress.

In the beginning, when an idea is conceived for an academic publication, there is always a question whether a journal edited by faculty and published by an independent printer could succeed. This third volume is both the sign and the result of the Journal of American Academy of Business, Cambridge's coming of age. The Editorial Board members took this opportunity to thank each and every reader and contributor for their devoted loyalty to JAABC. It is because of this relationship based on loyalty that they began in 2003 and continued to the present with positive momentum that stemmed from a stellar 2002.

Changes in the Meanings of Words

It was in the year 1666, a malefic fire swept through London, burning half of the city and destroying one third of St. Paul's Cathedral. Sir Christopher Wren, the original architect of the Cathedral, was commissioned to restore this much-loved, much-admired edifice to its former glory. It took thirty-six years to finish the careful restoration. On the inaugural day, Queen Anne, the reigning monarch, visited the Cathedral and told Wren that his work was *awful, artificial* and *amusing.* Sir Wren enjoyed the royal complements, because in those days, "awful" meant awe-inspiring, "artificial" meant artistic, and "amusing," meant amazing! When a nation's language changes and if the people wish to keep up with them, such an undertaking requires flexibility and adaptability.

The change in the meanings of Queen Anne's words was well over three hundred years ago. Today, the older flattering meanings of awful, artificial, and amusing are virtually obsolete from popular currency. While the evolutionary changes in a language is at a glacial pace, changes in the thought and practice of business have a revolutionary speed. Overnight, old concepts are discarded and new ones are adopted, instead. A case in point is relationship marketing. Not too long ago, management's focus was on increasing sales volume by finding new customers; now, the drive is for building relationships through customer satisfaction for repeat business. The market expansionist ideology has given way to the strategy of gravitating on the existing market (i.e., customers).

The meaning of words change, the viability of concepts change, the applicability of methods change. When new concepts, new methods, and new philosophies emerge in business, the Editorial Board strives to keep up with the rapid river of change. As Heraclitus said, "One cannot step twice in the same river." We, members of the Editorial Board, need to use speed and vision in grappling with change. We believe strongly in "upward mobility," which is a form of change not for status though, but for our readers and contributors' increased satisfaction upon which to build a relationship.

The Success Story of a Journal Based on Relationships

While quantum leaps are better in improving quality of service, it is too expensive in the journal publishing business. Therefore, given limited resources, the publisher of JAABC has resorted to incremental changes in the quality of the periodical. Compared to the pervious issues, they have increased the number and variety of articles. Compared to our pervious issues, they began to use slick paper for better appearance, easier readability, and enhanced durability.

Furthermore, in addition to being accepted last year by ABI/INFORM (ProQuest) for inclusion in their internationally acclaimed business database, in 2003 JAABC began to be listed in the Ninth Edition of Cabell's Directories! Slowly, but surely, JAABC was getting the well-deserved recognition. In this way, they attempt to build a relationship with their readers and contributors through initiating changes to improve the Journal.

Building Relationships through Customer Satisfaction

Building relationships means that the Editorial Board should be sensitive to change such as in relationship marketing. This is a relatively new strategy for management that entails forging long-term partnerships with customers. Companies build relationships with customers by offering value and providing customer satisfaction. They are rewarded with repeat sales and referrals that lead to increases in sales and profits. The success of JAABC is also rooted in relationship marketing. Not only do they offer their readers with timely and provocative articles, but they also organize conferences to provide them with a forum for the exchange of emerging concepts in business.

Embracing change for the sake of change, of course, could backfire. However, change for increasing the satisfaction of loyal customers such as the readers of JAABC would be beneficial. Therefore, the Editorial Board members of JAABC have exercised the elasticity of the mind to keep abreast of changes, by being responsive to changes, and by initiating changes. They bow to the inevitability of change since they are all staunchly convinced that the only constant in business is change. Resting on one's laurels is tantamount to ignoring the potentialities of change. The Editorial Board has been, therefore, all ears to hear your ideas for further improving the Journal since they valued their relationship with the readers and contributors of the Journal.

The publishers of JAABC have the vision to move beyond the parameters of providing a good periodical. Consonant with the spirit of relationship marketing, through incremental improvements, their goal is to establish one of the premier academic journals in business. Endowed with boundless energy and determination, they have already made headway in achieving their goal. The end result of building a relationship is loyalty, which is considered the queen of all virtues. For the Editorial Board of JAABC, loyalty is also the quintessential key to success in all human endeavor and enterprise.

CHAPTER
FOURTEEN

Commitment:
The Quintessential Element
in Success

O NE UNEQUIVOCAL BASIS for success is commitment. Committing ourselves to a goal or a purpose, and remaining dedicated despite distractions or challenges, can yield lasting rewards. According to The American Heritage College Dictionary "commitment is 3. The state of being bound emotionally or intellectually to someone or something." So, please indulge me with a few moments of your time to exemplify as to how commitment brought the best for a not-for-profit organization.

How Commitment Made a Journal a Smashing Success

Therefore, let me take an example from the real world to explain how by virtue of commitment the Journal of American Academy (JAABC) flourished in a short period of time. The staff members of the Journal of American Academy of Business, Cambridge, are committed to providing a forum to academicians and professionals for the exchange of ideas, views, theories, and paradigms pertaining to the conduct of business of local, international, and global scope.

In addition to the major project of producing the Journal, the energetic and future oriented publisher organizes various national and international conferences, such as in Istanbul, Turkey; Miami, Florida; Honolulu, Hawaii; and London, England. Programs such as these are designed to foster the interchange of new ideas and cutting-edge methods of doing business in this ever changing era of globalization.

JAABC's Board of Directors' commitment to quality shines through the past and the present issues of the Journal. The scholarship contained in these articles is testimony to the fact that the Board Members, Reviewers, and Editorial Advisory Board's policy is to select the best manuscripts for publication. Recently, the Journal was accepted by ABI/INFORM for inclusion in their internationally acclaimed business database, ABI/INFORM. This is another example of commitment to ever improve.

JAABC's commitment to timely articles is transparent. A mere glance at the table of contents of any issue of the Journal would show one that the Journal consciously addresses current issues and problems from various interdisciplinary perspectives.

As many readers know well, most academic journals belly up within several years of their inception because of lack of funds. Some may find the registration fee a bit steep for publishing an article, but because of the commitment to the readers or contributors, thus far has had no problems in raising funds. The whole idea is to make sure that JAABC gets enough funds for the continuity of the Journal. A dead stallion, although once beautiful, is of no use to anyone's need for a race to reach a destination! It goes without saying that the right funding is essential to the survival. Without proper survival, meaningful growth would be improbable, if not impossible.

If nations, armies, corporations, and individuals have succeeded, it is because of a deep and unwavering collective commitment of some individuals to ideas and ideals.

How a Nation Was Catapulted into Success

For an historical example, let's look at a small island kingdom. When the island committed itself to Sir Walter Raleigh's ideal as captured in the statement "Whosoever commands the sea commends the trade; whosoever commends the trade of the world commends the riches of the world, and consequently the world itself"(in 16th century), it performed legendary feats in global politics and trade. Commitment to this ideal made England by 1815 the vastest and the richest empire the world has ever known!

Commitment is the Key to Success in Any Endeavor

Commitment is the cornerstone of all endeavors, be it in business or in academic activities. The publisher of JAABC through his firm commitment, through all types of weather, to a clear and uncompromising goals and objectives are paying the highest rewards now. JAABC has been flourishing for over a decade now including the conferences. Both of these activities also reflect the organizers' commitment to perseverance.

The publisher of JAABC is the personification of the spirit of commitment. His unswerving dedication and tireless efforts to produce the nationally accepted Journal as well as to organize internationally attended conferences are the direct result of his commitment.

It is because of commitment of certain people behind the scenes who have been the "movers and the shakers" and who have stood steadfast to face the challenges, made a resounding popular success of JAABC. This Journal has become a premier publication wherein the domestic and the international erudite community can find business related articles of theoretical and practical orientation.

As the JAABC's example illustrated, commitment is the key ingredient in any endeavor, be it small or large, by an individual or a nation which inevitably leads to success.

CHAPTER FIFTEEN

Reverse Evolution: A New Frontier in Science

IT IS FAST becoming a truism that science enjoys limitless horizons of discoveries. Today, there is a new kid on the scientific block which attempts to bring the past to the present and future without any hypnosis or magic.

A case in point is Reverse Evolution. Basically, the attempt here is to how to hatch a dinosaur.

The Creation of Atavisms

Scientists are working on how to create atavisms in the laboratory. (Atavisms are long-extinct creatures, echoes of evolution past, which occasionally emerge in real life. Such rare cases of individuals born with characteristic features of their evolutionary antecedents are well documented).

For example, whales are sometimes born with appendages reminiscent of hind legs. Human babies sometimes enter the world with fur, extra nipples, and in very rare cases with a true tail behind their butts.

Over the past several decades, paleontologists such as Jack Horner have found overwhelming evidence to prove that modern birds are the descendants of dinosaurs. The similarities include everything from the way they lay eggs in nests to the details of their bone anatomy. In fact, there are so many similarities that most scientists now agree that birds actually are dinosaurs, most closely related to two-legged meat-eating theropods like Tyrannosaurus Rex and velociraptor.

The Idea of Reverse Evolution

One of the leading proponents of reverse evolution is Jack Horner. His plan is to start off "creating" experimental atavism in the laboratory. Fundamentally, his idea is to activate enough ancestral characteristics in a chicken, and one would end up with something close to the ancestor to be a certain kind of a dinosaur.

Horner's method to reverse evolve a dinosaur is not how most people imagine how the T. Rex making a comeback. That scientific premise was held by Michael Crichton's ***Jurassic Park.*** That is to say, bloodsucking insects trapped in prehistoric amber could contain enough dinosaur DNA for researchers to clone these big animals.

At first, Horner embraced this notion and threw himself in studying its feasibility. He ultimately gave up the idea for he concluded that DNA breaks down too fast in amber and in bones no matter how well preserved the trapped animal is. In other words, he concluded that dinosaur cloning was not feasible yet nor it was in the grasp of scientists.

As a result, Horner began to study developmental biology in order to create a dinosaur out of a chicken.

Already, scientists other than Jack Horner have found tantalizing clues at least some ancient dinosaur characteristics can be reactivated.

The Reactivation of "Chickenosaurus"

Horner firmly believes that dinosaur characteristics reactivation depends on a few breakthroughs in developmental biology and genetics and some chicken eggs. He claims that we would take our chicken, modify it, and make a "chickenosaurus."

Should Horner's dream be realized, the pet industry would explode with a runaway demand for small dinosaur pets for children including a lot of grownups.

Farmers would be delighted to sell not just chicken eggs, but very big and nutritious "dinosaur eggs" for breakfast.

McDonald's restaurant would be the first to capitalize on this discovery by offering a new breakfast meal dubbed as "McDosaurus."

Some Good and Bad Implications

Like screeching peacocks roaming free in some neighborhoods, runaway dinosaurs would canvas the neighborhood streets in search of food and would find our domestic pets (e.g., cats and dogs) prey for dinner!

Dinosaurs' comeback could create problems to mankind. Because of the carnivorous dinosaurs, the Homo sapiens would return to live in the trees for safety. Moreover, the vegetarian dinosaurs would turn the globe into a vast desert to satisfy their voracious appetite for whatever is green.

After researching this avant-garde topic, I realized that scientists speak of "creating this and that." I thought only God and the Pope could "create," and no other mortals. However, science has a way to make us stop and think as being another earthly god shaping and guiding our daily and future life.

CHAPTER SIXTEEN

Social Entrepreneurship: Sustainable Solutions to Societal Problems

THROUGHOUT THE AGES, empires have risen and empires have fallen, only one thing has remained constant, namely lack of adequate social services. In modern times, the British Empire has been so vast and so rich that people used to say "The sun will never set on the British Empire." Even the richest and the biggest empires of them all the likes of the Hellenic Empire, the Roman Empire, and the Ottoman

Empire could not muster up enough resources to go around to the segments of the population stricken with poverty. To this day, England has neighborhoods where the inhabitants have to use community baths and bathrooms.

The Burden of Charity Was on the Nobility

In the olden days, it was "noblesse oblige;" the burden was on nobility to take care of the poor to some extent. The rest was picked up by the church, the mosque, the synagogue, or the temple which used social services to maintain or attract new members and converts. Charitable actions or disposition toward those classes perceived as low or lacking necessities of life were not well organized. Although entrepreneurs existed for centuries, social entrepreneurs are the recent breeds of the 20th century. For example, In 1889, Jane Addams founded Hull-House, a social settlement to improve conditions in a poor immigrant neighborhood in Chicago; Maria Montessori, the first female physician in Italy, began working with children in 1906 and created a revolutionary education method that supports each individual child's unique development; and Muhammad Yunus introduced Grameen Bank (village bank) in Bangladesh in 1976 to offer "micro" loans to help the village toward economic self-sufficiency by producing products on their own. Yunus's model has been emulated in 58 countries. For his contributions to alleviate world poverty, Yunus won the Nobel Prize for Peace in 2006.

The Age of Charity Organizations

Today, the United States is the richest country in the world and yet it has to depend on not-for-profit organizations, charities, various religious institutions, and social entrepreneurs to meet the growing demand for social services in an environment where the divide between the haves and have-nots seems never possible to bridge. However, there is a strong trend in harnessing individual and collective efforts to provide for the needy ones.

These are the times of challenge and change. Our business students will define the destiny of social entrepreneurship in our society. They are our premier pool of our future activists and agents to cause change, to improve society, to render the world a better place to live. The only true boundaries in innovation and creativity are the ones that are self-imposed.

Evolutionary and revolutionary forces have always characterized life as being constantly changing on the planet Earth. That the world of the entrepreneur has changed should come as no surprise. The landscape of business and the economy is no longer the same. It has been proclaimed that the new breed of entrepreneurs has initiated a new era by ending the Brontosaurus capitalism of the 19th century. Some predicted a few years ago that "The Internet has been and will continue to be a testing ground, a venue for new ideas and new technologies." Changes always bring along challenges and opportunities for those who live in the future with their eyes glued on the frontiers of innovation and creativity. The history of inventions must have witnessed many dawns, creating a different environment for the entrepreneur to bob and weave his or her way through events, objects, and people.

The Rise of Social Entrepreneurship

There is a growing trend of an enterprise in the domain of entrepreneurship which is attracting a considerable attention from the academics, business professionals, movie stars, and politicians. The new kid on the block is called social entrepreneurship (SE) who has a big heart and a bright future for "midwifering" of essentials needed by the have-nots of society.

The concept of social entrepreneurship centers on the "win-win" strategy of doing well financially while doing "good" for society. SE has been accepted as being beneficial in reducing hunger and poverty, promoting good governance, advancing sustainable development, while creating wealth for investors. As the practice of SE proliferated, it has branched out into a couple of dozens of challenging areas where society needed much assistance such as in filling the gap in social services. A thorough review of the literature has shown that there are some areas which are already being used for social assistance.

Main Areas of Social Entrepreneurship

Altogether twenty mutually exclusive major areas of SE specialization opportunities have been identified which are listed below, each area with a brief operational definition for further clarification:

➤ Digital Divide (The world is divided into people who do and people who do not have access to modern information technologies or lack the education to handle them).

➤ Affordable Housing (Housing in which costs are at a level that does not threaten other basic needs and represents a reasonable proportion of an individual's overall income).

➤ Alternative Energy (Energy derived from nontraditional sources, such as compressed natural gas, solar, hydroelectric, or wind and usually environmentally sound as opposed to fossil fuels).

➤ Environmental Industry (Involves management of solid, hazardous, and medical waste, including the manufacturing, distributing, and servicing of waste equipment and offering related pollution-prevention services).

➤ Agriculture (The science, art and business of cultivating the soil, producing crops, and raising livestock).

➤ Water (The science and technology used in water purification and/or drinking water treatment).

➤ Forestry (The science, art and practice of managing and using trees, forests and their associated resources).

➤ Nutrition (The study of food ingredients and liquids needed to maintain a healthy life).

➤ Health (The study, art and practice of preventing disease and maintaining health).

➤ Medical (To help find equipment, methods, and approaches to facilitate the practice of medicine).

➤ Bioethics (Ethical and equitable methods of dispensing of biological sciences and medicine).

➤ Education (Bringing up new methods and making them available for teaching and learning specific skills).

➤ Literacy (Providing opportunities to improve one's ability to read, write, and comprehend).

➤ Diversity (Sensitivity training of others to become knowledgeable about the diverse dimensions of people in terms of age, gender, race, ethnicity, ability, and religion).

➤ Multiculturalism (To provide training on how to deal with different cultures and how to learn to get along with one another with mutual respect).

➤ Entrepreneurial Opportunities for the Disabled (To offer opportunities for business ventures to the disabled in terms of venture capital, training, employment programs, and relevant facilities tailored to people with disabilities).

➤ Human Rights (To train in and advocate the basic standards of civilization without which people cannot live in dignity such as the freedom of speech and expression).

➤ Social Services (To help develop and/or provide human services, generally provided by the government, or various agencies to improve people's standard of living).

➤ Funding the Social Entrepreneurial Venture (Helping others by securing and making available venture capital for social entrepreneurship, dealing with realistic, affordable, profitable projects beneficial for society).

➤ Corporate Social Responsibility/Performance (corporate citizenship extends the boundaries of the company beyond its shareholders toward society at large for enhancing broader societal goals, policies, and programs and not merely to benefit a more restricted number of shareholders and/or stakeholders).

The above areas are challenging opportunities for SE to explore and find creative ways to be of service to that segment of society which sorely needs external assistance. Obviously, these areas in SE would also serve as possible topics for us to research. My colleagues and I salute SE for its noble mission and vision to find sustainable solutions to help alleviate the privations of our less lucky fellow members of world society.

CHAPTER SEVENTEEN

Corruption: The Ultimate Cancer

L AST YEAR THE purchase of a summer house in Eastern Europe subjected me to a number of corrupt activities in my attempt to consummate the deal within four months. I was appalled at how government officials expected bribery to get the necessary paper work done. As a bribe giver, I was as guilty as the bribe receiver, but I had to get the title of the property as soon as possible and return to the States to my work.

The Nature of Corruption

Corruption is universal. The general consensus has been that it exists in all countries. In some countries, though, the practice is widespread and deeply entrenched in the daily lives of the society. As a result, it has become a major social issue around the world. The consequences of corruption are taking a heavy toll not only on the indignation of righteous citizens, but that it is also devitalizing, if not crippling, national economies from forward progress.

Lack of transparency, accountability and consistency coupled with weak legislative and judicial systems provide fertile ground for the growth of bribery-based activities in some countries. In addition to creating an underground parasitic economy and causing high social cost, corruption is creating adverse consequences on personal and national investment, the government budget, and on economic reforms of a nation.

In this chapter, corruption is defined as the use of public office for private gain, or use of official position, rank or status by an office bearer for his or her own personal benefit. Based on this definition, examples of corrupt behavior would consist of bribery, extortion, fraud, embezzlement, nepotism, cronyism, appropriation of public assets and property for private use, and influence peddling.

How to Get Rid of Corruption in the Individual

Many articles have been written to point the ills of corruption, but only a few have offered a workable solution to shaking off the addiction. Drawing upon psychological concepts and theories, an attempt is made here to suggest a method out of the quagmire of corruption from an individual's standpoint.

What are the cures for corruption? As there is no sure cure for cancer, likewise there is not any quick fix for corruption. Often we blame our leaders for corruption without really pinning the problem on the people, the citizenry of the country. No national leader, such as the president of a country, has the power to stop corruption in a country. There would not be enough prisons available to accommodate millions of corrupt people in his or her nation. It is the moral obligation of each and every citizen to realize that he or she is harming their country through the practice of corruption out of selfish personal reasons.

Often one hears in the former Soviet Republics and in Eastern Europe that the practice of corruption is left over from the Soviet era and that nothing could be done to eradicate the practice. The world community has finally realized its debilitating effect on progress. This greater recognition that corruption can produce serious adverse effects on the development of an economy has sparked alarm among developing countries. In a recent survey of 150 high level officials from 60 third world counties, the respondents ranked public sector corruption as the most severe obstacle confronting their economic development prospects.

What Are the Basis of Corruption

Before suggesting a method, we should first understand how corrupt behavior is formed. Fundamentally, our character is the grand total of our habits. The Greeks pondered on this issue at length thousands of years ago. Aristotle (384 BC-322 BC), however, summed it up by stating that "We are what we repeatedly do. Excellence, then, is not an act, but a habit." By the same token, if one engages in corrupt behavior repeatedly, then he or she is corrupt. Moreover, we can say that corrupt behavior is not an act, but a habit. If an official accepts bribery repeatedly, we can say that that corrupt behavior is not an act, but a habit.

As it has been expressed in the following conventional wisdom, "Sow a thought, reap an action; sow an action, reap a habit; sow a habit, reap a character; sow a character, reap a destiny," so goes the maxim.

From a psychological perspective, habits are powerful factors in our daily lives. Because they are consistent, often unconscious patterns of behavior; they constantly, almost daily, express our character and produce our effectiveness or ineffectiveness.

As Horace Mann (1796-1859), the great leader and educator, once said, "Habits are like a cable.

We weave a strand of it every day and soon it cannot be broken." I do not agree with the last statement, though. Although corrupt behavior is the ultimate cancer of a society, it is curable. The difficulty of curing this affliction lies in the fact that corruption has been deeply embedded into the culture of a society. It would be hard, but not impossible, to change that culture.

Habits Are Reversible

Based on scientific psychological evidence, habits can be broken. Habits can be learned and they can be unlearned with great difficulty though. To unlearn a habit, one has to go through a regimented process and an unwavering commitment.

To quit smoking, for example, one has to go through a long process of withdrawal symptoms. Habits in general require tremendous patience and resolve most people fail to realize or admit. Breaking deeply imbedded habitual tendencies such as procrastination, impatience, criticalness, or selfishness in accepting bribery that violate basic principles of human effectiveness involves more than a little willpower and a few minor changes in our lives. It involves deep commitment and genuine love of one's own country.

We can define a habit as the intersection of knowledge, skill, and desire. Knowledge is the theoretical paradigm, the "what" to do and the "why". Skill is the "how" to do. And desire is the motivation, the "want" to do. In order to make something a habit in our lives, we have to have all of the three elements. Likewise, in order to unlearn or break a habit, we would also need to have the previous three elements present.

For example, I may be ineffective in my interactions with my work associates or my spouse because I constantly tell them what I think, but I never really listen to them. Unless I search out correct principles of human interaction, I may not even know I need to listen.

Even if I do know that in order to interact effectively with others, I really need to listen to them. I may not have the skill. I may not know how to really listen deeply to another human being.

But knowing I need to listen and knowing how to listen is not enough. Unless I want to listen, unless I have the desire, it won't be a habit in my life. Creating a habit requires work in all three dimensions.

Summation of Corruption in a Descriptive Formula

Thus, corrupt behavior can be unlearned through the following descriptive formula: UC = f(KxSxM); UC= this means to unlearn corrupt behavior or to break the habit, it is the function of the intersection of K= knowledge, S= skills, and M=motivation.

Very briefly, Mr. or Ms. Corrupt citizen should have Knowledge of what the problem is or what are the consequences of corrupt behavior, etc. the citizen should realize that the stakes are high for himself as well as for his nation. For himself, he would be running the risk of losing his job and being sent to jail. For his nation, corruption means devitalizing the economy for many years to come. Every corrupt act would be akin to driving a nail in the coffin of one's country's economy.

As for Skills, the citizen should seek ways to unlearn the habit. Post notes on your desk which remind to say "Nyet" or "Nein" to bribery. Reward yourself with something you like every time you refuse to accept any bribery. Try to reinforce the will power in you to avoid engaging in corrupt behavior. Every individual has his or her way of using personal skills to deal with the unlearning of a habit.

Feeling of Self-Respect from Abstaining from Corruption

Finally, for Motivation, the citizen should visualize himself as being respected as a righteous person, by his fellow citizens, by his family and friends. A self-respect emanating from the realization of being a true citizen contributing positively to the progress of his country by not engaging in corrupt behavior would serve as the powerful psychological inner satisfaction of doing something right. It is the duty of every citizen to consider corruption as the ultimate cancer gnawing at the development of their country's bid for progress. Anyone engaging in corruption would be eventually spitting in the wind.

We should all salute and wish the best of luck to those individuals and leaders who are doing something to get rid of the cancer of corruption in their countries as well as in international business.

CHAPTER EIGHTEEN

Digital Deprivation Effects on Entrepreneurship

THE MUCH-DEPLORED digital divide phenomenon has been well documented for over several decades now. The disparity of the application of new technologies between industrialized nations and emerging markets has been noted to have been widening. While the entrepreneur in the United States does not live in the lap of digital luxury, the developing nations' counterpart is lopsidedly deprived of the latest cutting-edge technologies. Technologically challenged entrepreneurs, however, are found everywhere regardless of the presence or absence of digital innovations.

The Digital Divide

Some of the observers of the information technology scene state that entrepreneurs in "have-not" countries, such as in Latin nations, Middle East, and Eastern Europe, are thwarted in realizing their innovative ideas. Unfortunately, there is no evidence to support this contention. In the absence of empirical evidence, one can point out that the paucity of technologies may not hamper nor dampen the entrepreneurial spirit and action. Brazil, for example, lacks basic digital technologies and yet it is cited as number one Latin nation in the world which has the most entrepreneurial activities.

While entrepreneurs all over the world dream impossible dreams, how do entrepreneurs in Latin America or in Eastern Europe fare in transforming ideas into successful business against an increasing chasm in the digital divide? One way to settle these conflicting reports is to conduct a study in a certain Latin American country, say in Columbia or in Argentina, to determine whether the digital deprivation affects adversely the entrepreneurial initiative and action.

The current issue hinges on the "haves" and "have-nots" nations in terms of the latest information technologies. Highly industrialized nations enjoy extensive markets both domestically and internationally, while most Latin American countries are limited to small domestic or regional markets. Therefore, the digital divide may not be a factor in Latin American entrepreneurship.

Questionable Assumptions on the Effect of Digital Divide

The tendency has been to embrace new technologies wholesale. Imagine all the plethora of tools which have been invented for home gardening, yet in a field in Africa or in one's backyard one can cultivate a small vegetable garden with a single shovel.

The assumption has been that because the developing nations' entrepreneurs lack the latest information technologies, they are deprived from realizing their dreams. There is no empirical evidence to back that position. At best, it is an untested proposition. Technology and creativity, like poverty and happiness, do not necessarily go in the same direction. Just because of lack of new information technologies, it would be erroneous to assume that there is a dearth of entrepreneurial initiatives and innovations. The media and the academics have created an imagined entrepreneurial disability to have existed due to unavailability of new technologies, say in Latin American countries.

We cannot superimpose our way of doing business over that of our Latin American neighbors. The media write with metaphysical certitude that the have-nots are at a great disadvantage. Jumping to such a conclusion may distort reality. Take, for example, the electronic calculator. While it is a useful instrument to help expedite computation, its absence does not necessarily mean that the entrepreneur cannot do expert mathematical calculations manually.

Latin America under
Digital Deprivation

Latin America, for example, is characterized by small markets and usually entrepreneurs there begin from a small nucleus in a neighborhood with a limited trading area. Therefore, it might be that sophisticated information technologies may not be essential in building a successful enterprise. Perhaps Latin American entrepreneurs resort to heuristic rather than to digital methods. Entrepreneurs are the backbone of a nation's business and industry; and it may very well be that it is quite advanced and widely practiced in a have-not Latin nation. Many Latin American nations, such as Mexico, Brazil, and Argentina, have been known to have been rapidly industrialized in the Western Hemisphere without having the luxury of the latest digital technologies.

Unless we have definitive evidence to the contrary, we should not adhere to the myth that technology is the essential ingredient for entrepreneurial activities. While numerous articles have been written on the issue of the digital divide, none has truly addressed the question of whether inadequate information technologies have had adverse effects on entrepreneurship. The challenge is to conduct a scientific study on the effect of digital deprivation on developing nations' entrepreneurs' behavior (not perception).

CHAPTER NINETEEN

Global Corporate Diplomacy: The Strategy for Survival

ONE OF SHAKESPEARE'S characters has bemoaned in a soliloquy this way: "Reputation, reputation, reputation! O! I have lost my reputation. I have lost the immortal part of myself, and what remains is bestial." The American corporations are being written out of good reputation by most people around the world. On account of the United States' foreign policies, multinational corporations' "remaining" perceived "bestiality" (i.e., brutishness, depravedness) is hurting their competitive position in the global marketplace.

The United States' Foreign Policy on Corporations

The repercussions of foreign relation decisions of the United States government have had far-reaching implications for the corporate world pursuing a globalization strategy.

Although negative perceptions of the United States began two decades ago to materialize, it reached maturity with the advent of George W. Bush administration. President Bush's controversial, sometimes-belligerent, but often perceived dubious foreign policies along with encroaching globalization inroads by such brands as McDonald's and Starbucks have created anti-Americanism. Such a debilitating sentiment is being shared by many people all over the world.

Europe and South America have lined up with the Middle East against the United States, especially for the handling of the Arab-Israeli conflict and for the protracted war in Iraq and Afghanistan.

On September 14, 2006, leaders of the nonaligned nations met in Havana, Cuba to bash the United States and solidify their opposition against its foreign policies. President Hugo Chavez of Venezuela went as far to call President Bush "The Devil."

As Alan Greenspan put it in a speech in 2000: "In today's world, where ideas are increasingly displacing the physical [products] in the production of economic value, competition for reputation becomes a significant driving force, propelling our economy forward." He went on to say that "Manufactured goods often can be evaluated before the completion of a transaction. Service providers, on the other hand, usually can offer only their reputations."

Dependence on Reputation and Image for Success

Most U.S. corporations are in service-based industries. Thus, the digital economy will make it imperative for companies to lean heavily on their good reputation and image to be successful in this world of hypercompetitive marketplace.

Deception on the Capital Hill and corruption in the U.S. lobbying activities have eventually alienated the rest of the world and caused the downfall trend of the U.S. multinational corporations' reputation in the world business arena.

The war in Afghanistan and the invasion of Iraq have had far-reaching ripple effects beyond politics. The major repercussions of Bush administration's foreign policies are also impacting on the United States international business. How people of other nations think about the United States is spilling over into the area of business and commerce. According to a recent poll by GMI, a market-research company, 80 percent of European and Canadian consumers, for example, distrust the U.S. government. Gail Dutton in a 2005 article titled "Grassroots Diplomacy," published in From Across the Board, maintains that "Even if your company didn't launch the Iraq war – or make any other policy decision of the past few decades – those consumers aren't separating corporate and government interests . . ."

Nearly 50 percent of those surveyed indicated lack of trust toward American corporations. Thus, United States corporations can no longer separate themselves from the foreign policy decisions of their government. In the past, politics was viewed as separate from business. More and more, multinational corporations are finding the two to be inextricably linked in the perception of people around the globe.

In addition to carrying out foreign policies based on half-truths and deception, the recent government and business scandals, such as the Enron debacle, have eroded the credibility of the United States management practices.

Western Europeans, for example, have sublimated their ire on the United States businesspeople for being the nearest representative of the government they hate. Stuart Crainer and Des Dearlove in an 2005 article titled ("Is U.S. Business Losing Europe?" have argued that such a tenuous situation calls for corporations to engage in their own grassroots diplomacy to place themselves on a different level than their own belligerent government. Many executives believe that in foreign affairs, corporate diplomacy would be as essential as political diplomacy.

The Devil Incarnate

One would shedder at what the Arab world thinks of the United States. Most Moslem and Arab nations consider the United States to be the devil incarnate. That is why it is easy for the extremists to find suicide bombers to blow up the Americans and their allies and to create death and destruction without differentiating between the innocent people and the unscrupulous politicians who are allegedly responsible for devising and executing foreign policies perceived as questionable based on subjective, inequitable rationale.

The U.S. war-subsidized economy is fairly strong now, but after the conflict is over all the repercussions of the ill-conceived foreign policies may come home to roost. Once the level of standard of living of the average U.S. citizens deteriorates, the raw realization of the damaging U.S. diplomacy will come to haunt millions of people who have been silent about the corruption in the government.

We are not taking sides with either the Republicans or the Democrats. However, we are deeply concerned about the United States global business and commerce. It is our responsibility to safeguard the reputation and image of the United States corporations in the perception of the world for it involves our professions, singly or collectively, in Accountancy, Finance, Management, Marketing, Information Systems, and the rest. Lack of credibility and trustworthiness create diffidence in international business and marketing. The result would be a devastating blow to the United States economy.

The Unfair Victimization of U.S. Corporations

In the landscape of a fiercely competitive global business, the survival of the positive image of the U.S. corporations is at stake. For all of our colleagues – regardless of their political orientation – we challenge them to discuss, debate, and write as to how the United States corporations could employ a separate diplomacy to distance themselves from their government without hurting or reflecting negatively on the overall United States' government image.

That is truly the fundamental question to be addressed by our professional and academic colleagues in order to avoid the unfairly victimization of the U.S. corporations globally by being perceived guilty by association with their own government's foreign policies. Against the backdrop of increasing global competition, survival strategies for the U.S. corporations through innovative corporate diplomacy must be the responsibility of all its stakeholders who should serve as the loyal ambassadors of their benefactors.

CHAPTER TWENTY

Political Paradigm Shift during an Economic Crisis

As PLANS FOR economic recovery are being implemented and put on course, many conservatives are clamoring to sound off the alarm that the United States capitalism is being replaced by socialism. Therefore, the shift in the U.S. political paradigm is detrimental for all present citizens and future generations.

As President Barak Obama's administration is taking the helms of the government, the crescendo of fear is being transmitted in writing as well as through the talk show media that the "invisible hand" of the government is becoming visible in running the business of the nation. Some even are openly lamenting the demise of good old capitalism and the return of debilitating socialism. This sounds like a misconception at the least and disingenuous indictment at best.

A Synopsis of Socialism

The term "socialism" made its debut in the 1830s in England and France. Then as now, this political paradigm covered a wide variety of opinions and methods to remedy the ills of society. However, the fundamental proposition of socialism has always gravitated to the idea that production is a function of all members of society and that the rewards of production are therefore to be utilized by all.

In 1840, the French radical Louis Blanc (1811-82) pronounced the classic formulation of socialism by stating "From each according to his abilities, to each according to his needs." The roots of modern socialism goes back to the Enlightenment's concern for human progress in response to social problems and economic inequities of the people caused by the rise of industrialism.

The two men who have been considered to be the founding fathers of modern socialism are the French social scientist Claude-Henri de Rouvroy Saint-Simon (1760-1825) and the British cotton master and reformer Robert Owen (1771-1858).

After analyzing the nature of industrial society and the relationships between prevailing systems of belief and the structure of societies, Saint-Simon concluded that liberalism was too wedded to individual rights to be able to ameliorate the condition of the working classes and to give producers of wealth, and the laborers their just rewards. His approach would only be possible if the great landowners and the clergy were replaced by a new class of scientists, intellectuals, artists, and producers all united with concerted effort to direct activity toward the common good.

Owen's Cooperative Communities

Owen, on the other hand, was more practical. At his cotton mills at New Lanark in Scotland and at New Harmony in Indiana, he established cooperative communities and sought to put into practice his vision of a society in which ownership and control of the means of production were communal. Like Saint-Simon, Owen had arrived at the conclusion that a new society based on cooperative socialist principles could be created only on the base of a reinvigorated "Christian morality." (He should have also included, Jewish and Islamic morality for these religions too adhere to moral codes of conduct).

In the past 250, violent upheavals and sometimes peaceful evolutions have brought forth great change to the political scene in the world. The choices were always either or: Communism or capitalism, socialism or capitalism, and communism or socialism. The choices were distinct and binary.

From the dawn of civilization, every government had two major tasks: to defend its citizens from foreign attack, and to give them a legal framework to enable them live together harmoniously.

Yet, in the 21st century, governments are expected to do more than the two proverbial tasks: building and maintaining national infrastructure, running schools, hospitals, prisons, welfare, and regulating business and commerce to name a few.

Today's political systems are a synthesis of old ideas like democracy of city-states of ancient Greece had to merge with new concepts such as governments depend on the consent of the governed in a "social contract" as evolved in the 17th and 18th centuries. Such changes happened at a time of great political upheaval when people revolted against the authority of absolute monarchs.

Contingency Approach to Political Systems

Nothing seems to be in its pure form. Governments of the world apply contingency approach to political systems as the need arises. The United States of America is no exception. Deceit, lies, debasement, corruption, and, to crown all, greed have caused the present economic crisis. Whether a remedy to get the economy out of a slump is socialistic, communistic, or capitalistic would be acceptable as long as that system is temporary and is used as a contingency approach to urgently rectify the mistakes of the previous political system.

As an example to show the application of resources and methods different than their dedicated purpose, the U.S. Armed Forces are designed primarily for the protection of the nation against foreign aggression. Despite their specific designation of duty, they are sometimes called in to give a helping hand in a national disaster such as in the aftermath of a flood or an earthquake.

In the event of an emergency, when the ship is half sunk, any acceptable rescue effort should be applauded. In times of doom and gloom, only inaction should be criticized. The collapse of the U.S. economy was not due to capitalism, but to stark corruption in this system. The stimulus plan, the so-called socialistic spending spree, is destined to put the house of capitalism back in order by weathering the financial storm. If anyone group has a better remedy, should take center stage before the economy slides deeper into the recession.

Lessons from the Great Depression

Following the advice of John Maynard Keynes during the Great Depression of 1923, President Roosevelt initiated the Great Deal which was similar to President Obama's Stimulus Bill and was socialistic in nature and scope. After achieving the objectives of the Great Deal, the economy reverted to industrial capitalism as usual to the surprise of those steeped in tradition and old ideas.

One thing is for certain. If the wrong doers, such as the Wall Street banks, are being bailed out with the tax payers' money, those who have prudently and honestly carried out their responsibilities such as the production of the Journal of American Academy of Business, Cambridge (JAABC) should also be somehow compensated for their conscientious and ethical business practices for the editorial board worked diligently and creatively to contain costs and stay solvent in order to enable the regular production of a nationally and internationally acclaimed refereed Journal.

We should all heed to what Pastor Adrian Rogers once remarked "What one person receives without working for, another person must work for without receiving. The government cannot give to anybody anything that the government does not first take from somebody else . . . You cannot multiply wealth by dividing it." Without equitable distribution, it would be the end of any nation.

Pinning Our Hopes on President Obama

Let us hope that President Obama is neither a communist nor a socialist in the long haul. All he is doing right now applying a contingency approach to solve the mess created by his predecessor. Contrary to the contrarians, the stimulus package is a recovery program to resuscitate the ailing economy. Instead of criticizing the President for his swift actions to curb the nation's economy from spiraling into the black hole of depression, we should do all we can to encourage him to solve the current major problems and to do our share in rehabilitating the economy of this great nation.

CHAPTER TWENTY-ONE

The Love Industry Flourishes While the World Economy is in Shambles

ALTHOUGH ONE OFTEN hears the refrain that "love makes the world go around," little does the average person know about its many faces and its inner workings? Only recently the field of psychology has begun to scientifically study this vital area which is germane to all humans of all ages, genders, and walks of life.

Love is a complex phenomenon, encompassing friendship, passion, intimacy, and commitment including sex.

The Brain Circuits Studies on Love

Often, the topic has been shelved as mysterious beyond human comprehension. Recent research has also shed some light on the brain circuits involved in sex and love. Romantic love, which brings couples together, and maternal love, which binds mother and child, also has survival value. Not only love has essential survival value for the individual, but that it has the major function for the perpetuation of the species and the preservation of a nation.

The kaleidoscope of love runs the gamut of brotherly love, carnal love, filial love, and even homosexual love. In this chapter the focus is on romantic love, the passionate love between a man and a woman. At one time or another, virtually all of us have been involved with at least one of the different types of love, yet one takes it for granted such a valuable conception of love that one would know all about it when one is smitten by it.

The prevailing attitude is that women seem to know more about the modern conception of love than men do. Men are more preoccupied with sex and women are rather sticklers for love, which has created a lucrative industry in the entire world. It is mainly women's expectation that men observe and respect romantic love that the love industry is thriving in those hard economic times.

Love in the Medieval Times

In the medieval period, love was somewhat to be dreaded. Today, love is a welcome event in anyone's life. Such a fundamental change in attitude required many years of speculation, some research, and discoveries. For example of great importance in humankind's life is love. With the exception of courtly love, up to medieval times, love was considered to be an affliction not to speak about in public. Though love is complex in any age, romantic love (passionate relationship between a man and woman) during the medieval period consisted of two major elements that are entirely at odds with each other: suffering and pleasure (i.e., pain and lust!).

Today, psychology has recognized that "love" is essential for the wellbeing of the individual and for the survival of the species. Women have to feel fulfilled by bearing children and men by fathering them. Although psychologists in defining love differ, most agree that it consists of three major components: Passion, Intimacy, and Commitment. The presence of these ingredients between a man and a woman, render the relationship solid and enduring, graduating the relationship into a "Consummate Love" affair (the best of its kind). Studies have shown that all of these three elements foster bonding and a sense of belonging. Shakespeare, though, warns us in **A Midsummer Night's Dream** that "The course of true love never did run smooth."

Today Love is a Many-Splendored Thing

Despite all the pros and cons, love is hailed in modern times as "a many-splendored thing" as the popular song states and not as a "dreary thing" to keep it secret as a personal curse as believed in medieval times. Science is changing our perspectives on many issues as well too numerous to cite them here.

Without our insatiable appetite for romantic love, a large part of the mainstream economy would collapse. Luckily enough, the economic slump has not affected the love industry yet. Romance comes in many forms and most of them involve money in our modern world. It may be the initial winning and dinning or one of the more extravagant displays of affection required to put a relationship on a more permanent footing. There are so many various sources of information to act romantically nowadays. These are in the form of romantic entertainment books, films, music with which we put ideas into one another's heads.

Additionally, there are the various correctives such as flowers, chocolate, holidays that are often used to put relationships back on the right track when one of the partners feels the need to mend fences or to ingratiate himself or herself. Not every romantic gesture has to be expensive, but, taken together, it all adds up to a multi-billion dollar industry that, even in the current climate, is booming. For example, according to Forbes magazine, $17 billion were estimated to be spent by throbbing hearts on Valentine's Day alone in 2008.

Sales of Romantic books on the Rise

Sales of Mills and Boon romantic books have not faltered in today's shaky economy, either. It was during the Depression of the 1920s and 1930s that the sale of books really took off. To escape the gloomy days of the Depression, people took up reading light fiction of the romantic genre. Romantic books are expected with bullish certainty to continue to achieve record sales in today's dismal world economy.

It is not just the book publishers who are now profiting from the earning power of romance. The outlook also looks bright for the greeting card industry. In 2006, more than 2.87 billion cards were sold in the UK with a combined value of 1.306bn pounds. That is 48 cards per person. None of the countries in the world has an embedded culture of sending and receiving cards than the UK. Together there are about 800 card publishers operating around the country.

Chocolates are another tried and tested way of saying "I love you." Despite the downturn in the world economy, sales are booming. The seasonal and boxed chocolate market is now worth $1.246bn, a significant increase from $1.180bn in 2006, when a heavy downturn in the market was predicted.

As always, especially when most women around the globe have now become weight watchers and thus avoid sugar, flowers top even chocolate in terms of revenue. On Valentine's Day, of course, most of that trade is done in red roses, a flower associated with love, passion, and romance.

A more costly way of expressing love feelings is to buy a piece of jewellery to a loved one such as a diamond ring or a handmade, pure silver necklace with beautiful Sharma stones with matching earrings. The market is worth around $2.5bn, an 11 per cent increase over five years. The most successful jewelers are those offering high quality pieces in distinctive designs. Platinum engagement and wedding band of choice for young women are the real stars.

Gifts for Expression of Romance

Finally, gifts as an expression of romance is so deeply engraved on our nation's psyche, it looks as if the practice might withstand even the toughest of economic shocks. As we can glean from this write up that love is not exclusively in the domain of psychology. Because of its vast industry, business professionals, practitioners as well as academics, should have the motivation to study and write about love which makes the wheels of the economy roll even in hard times.

Against the backdrop of the ailing global economy, such a euphoric optimism seems to support the hypothesis that romantic love is not subject to the wild swings of the economy. Love seems to transcend the worst of economic downturns even in the form of a Depression in such unprecedented times. We should all, therefore, sing together the French song aptly titled: "Vive L'Amoure!" (Let us drink to our loved ones!).

CHAPTER TWENTY-TWO

Writing an Article vs. Writing a Book: Weighing the Wages

A FEW YEARS ago, when I was visiting one of my cousins overseas whom I had not seen for over half a century, during a conversation at the dinner table my cousin asked me a personal question in the presence of other guests. He asked me a point-blank question as to what books I had written when he found out that I had been a university professor for over twenty-five odd years. This type of question has made me uneasy in the past coming from lay friends and relatives, because it puts me in a defensive mode to explain as to why I do not have as many books to my credit as compared to journal articles. I told him that I had a couple of dozens of books and that one or two were forthcoming and that there was a long one in progress.

For fear that he might get a narrow impression about my contributions to my field, I hurriedly tried to explain to him that in my profession as a university professor, generally speaking, writing an article was considered rather scholarly for it advanced our understanding of scientific questions (i.e., it added to the existing body of knowledge in a certain area) and writing a book was somewhat looked upon to be a commercial endeavor. Besides, for tenure and promotion, papers and articles were expected from the candidates.

Articles Advance Knowledge

Having said that, not all articles advance the frontiers of knowledge, and not all books just rehash existing knowledge, either. There are books that have trail blazed new paths in science, such as the *Theory of Evolution* formalized by Charles Darwin, *The Principia: Mathematical Principles of Natural Philosophy* by Isaac Newton, and the *Meaning of Relativity* by Albert Einstein to cite a few.

The way my cousin had raised his eyebrows gave me the cue that he needed an elaboration; so I took an example of my research in Response Error (i.e., based on my proposal of a psychological theory as to why people misreport factual information).

Drawing upon one of my construct validity testing studies, I began to explain the manipulation of the independent variable, namely risk of being caught misrepresenting, on the experimental and control groups. Before finishing the methodology of the study, I sensed that either the topic was complicated, beyond him, or simply it was boring him. Thus, we switched the conversation to telling jokes after dinner. The dinner ended with a blast, though, with his hilarious jokes.

Upon my return from my long trip, I remembered the slight puzzlement on my cousin's face as I was attempting that night to explain why articles were preferred to books in my profession. Since he is a self-educated man with an inquisitive personality and an intelligent mind, I decided to share my thoughts on the merits of writing an article vs. writing a book with him and also with some of my younger colleagues who could very well be caught in a similar situation.

Writing an Article vs. Writing a book

In my view, writing a book on the automotive industry is one thing, writing a research paper or an article on a new method to save fuel is another thing. Comparatively, the second activity is a contribution to advance knowledge in an area of study such engineering. Therefore, I have spent the preponderance of my efforts and energies on doing research and writing papers or articles. The average person, unfortunately, does not seem to understand that.

I have also learned that one should avoid using "paper" to a lay person even if one were to explain that most articles had to go through the metamorphoses of being a research paper first, next it got presented at a conference, and then it saw the light by being published either in the conference proceedings or in a periodical as a journal article if it met certain quality standards. A "paper" has a bad connotation or is associated with a frivolous or ambiguous activity. The right word would evoke the right impression and, therefore, it would be better to use the word article as opposed to paper.

Unless one says he or she has five or more books to his or her credit, most lay persons do not consider that individual as having been productive in his or her profession. If, on the other hand, one were to say he or she had 100 articles to his or her credit, those who understand the workings of the academic world would be highly impressed. They would consider that individual as being a prolific researcher and writer, and even look upon him or her as a scientist in his or her field of study.

In fact, it is harder to be creative in advancing the boundaries of science than just sitting down and writing a book. To me writing a scholarly and creative article is more challenging and energy draining than writing a book. In writing a book one normally does not need to present other studies, while in a well-researched paper or article, the author has to go through the grueling task of reviewing the literature up to his or her study.

In an article, you have to be creative to advance knowledge. Ideally, there must be a gap to fill. Furthermore, to get an article accepted by a refereed journal, it has to go through the blind review process which may involve many revisions, while a manuscript for a book is only reviewed by one editor and the writer has the choice of either getting it published through a vanity publisher with no hassles by paying a fee or by a traditional publisher which would require more work to get a book published without a fee.

The Case of Identity Crisis

It seems that books and articles suffer from identity crisis. Many people do not perceive them as they really are. It is the same confusion with someone who has an MD degree compared to a Ph.D. degree. The vast majority of people show greater admiration and respect to an MD than to a Ph.D. holder. The fact of the matter is that a Ph.D. is the highest degree anyone can obtain from any university in the world. Yet, the average person does not know that.

Moreover, an MD is a practitioner (of physiology) just like a lawyer, a realtor, even a certified auto mechanic are practitioners, while a Ph.D. is a scientist of a chosen field and normally engages in basic research. A Ph.D. is trained to conduct scientific research while and MD is trained to dispense the knowledge gathered by a physiologist who is most likely a holder of a Ph.D. degree. As we all very well know, for a successful completion of a Ph.D. program, a candidate has to write a dissertation, an in-depth study of a specific topic or area, while such a demanding assignment is not required to get an MD degree.

Generally, books are held in higher esteem than articles, but upon a closer examination we see that articles have been underestimated unfairly and lopsidedly by most people. The correct designation is journal article. Journals are the place where researchers in the past recorded the findings or results of their studies by observation, survey, or experimentation.

Journals and Journal Articles

Aristotle (384-322 B.C., a Greek philosopher and scientist), is the father of scientific observation, who, for example, observed the behavior of the male and female cat fish in his personal living laboratory "aquarium" through which a natural stream flowed. Archimedes (287-212 B.C., a Greek mathematician) was the father of scientific experimentation. He used the research method in experimenting with several sizes of his invention of the water screw (to extract water from a lower level to a higher level for irrigation). Later on, Emile Durkheim (1858-1917, a French sociologist and philosopher) became the father of scientific survey research. He used the survey method in his studies of sociology of social solidarity and organic solidarity incidence in urban and rural areas of France. All three researchers used the journal to record their findings of what they were studying systematically. Hence now we have journals and journal articles containing the results of studies.

The activities of scientists remind us of another vital distinction between a book and a journal article. Generally speaking, books are based on secondary information or data which have already been collected and or published by others at different points in time. The validity (relevancy) and reliability (accuracy) of the secondary information or data are out of the control of the book writer. Journal articles are relatively based on primary information or data generated by the researcher for the study at hand. Primary information or data are current. Also the validity and reliability of the data are controlled by the researcher.

Benefits of Conferences

In comparison with books, there are many yearly conference opportunities for the presentation of research papers or articles. Normally, papers are written and then presented at the conference where the author would get input and insight from other colleagues. Another beneficial aspect of a conference for papers or articles is that it affords authors to travel nationally and internationally. The trip is usually funded by the institution with which they are affiliated. One seldom comes across a call for an academic conference where one can present a book. Most, if not all, universities do not grant scholarly and creative activity assigned (free) time for research nor do they provide student assistance if the professor proposes to write a book.

It should be pointed out that books are a source of social recognition, influence, and remuneration. There are books that make six figures in royalty per year, especially the popular text books used at institutions of higher learning. The wages of writing an article is scholarship, while the wages of writing a book is power, prestige, and money. Thus, scholarship is the single most important feature which renders an article superior over a book in general terms.

Admittedly, books are like golden ornaments in the landscape of one's resume which attract attention, whereas articles are the lackluster legacies of a researcher. In the real world, as compared to the academic environment, a person who writes a book is perceived to be more knowledgeable and even as an expert in a certain area. Hence the expression: "He or she wrote the book on . . ."

Within an academic environment, the wages or benefits of writing a scholarly article far outweighs the money a book would bring as is shown in Table 1.

For example, if one were to skim through the pages of an academic journal of, one would discover new ideas, methods, approaches, models, theories, paradigms, heuristic techniques, etc. all are proposed to either solve a problem or ameliorate a situation in various disciplines. That is what makes a scholarly journal an exciting kaleidoscope of articles contributing its share to the advancement of science and technology.

Table 1

Summary of Benefits of Writing a Paper or an Article vs. Writing a Book In an Academic Setting

Writing a Paper/Article	Writing a Book
➢ It is considered scholarly	➢ It is considered commercial
➢ It advances existing knowledge	➢ It rehashes existing knowledge
➢ It is usually based on primary information or data	➢ It is usually based on secondary information or data
➢ It goes through blind reviews	➢ It does not go through blind reviews
➢ It is usually published by not-for-profit organizations	➢ It is usually published by profit-oriented companies
➢ It is not often published by vanity journal outlets	➢ It is usually published by vanity book publishers
➢ It does not require marketing	➢ It does require marketing
➢ It has a narrow readership	➢ It has a wide readership
➢ It is ephemeral	➢ It has pass-along readership
➢ It is an essential input for books	➢ It nourishes on the findings of papers/articles
➢ It affords assigned time	➢ It affords no assigned time

➢ It provides for paid travel opportunities	➢ It does not provide for paid travel opportunities
➢ It is not a source of income	➢ It is a source of income
➢ It has low source credibility	➢ It has high source credibility
➢ *It is not a symbol of power & prestige* (for the lay person)	➢ It is a symbol of power & prestige *(for the lay person)*

In the final analysis, if I could not convince the reader of the superiority of the article over the book, then perhaps a compromising statement that both are essential ingredients in the education of the mind would hopefully be acceptable. However, let us not forget that books are normally based on the findings reported to the public in articles. Therefore, articles are the foundational sources which nourish the pages of books. So I would say to my colleagues, please keep on cranking out your articles for they pave the way to greener pastures in science and technology for improving the human lot on the planet Earth.

Most of my colleagues join me in upholding the principle that one should not judge the caliber of a professional primarily by the number of books he or she has written, but rather by giving equal, if not more, weight to the number of refereed articles the person has produced throughout his or her career.

So let us not make the mistake of respecting the article writer less than the book writer or visa versa. Even though perceptually books command higher source credibility, the least one could do is to respect them both equally as compared to the quasi-professional who has been sterile in his or her entire career despite many attempts at artificial insemination (through assigned time, travel funds, student assistance, lighter teaching load, etc.) and crosspollination or even crossbreeding (through opportunities to collaborate with other colleagues) by his or her frustrated Dean.

Coal is burning in these smokestacks.

Part III

Fundamental Issues Facing Society

CHAPTER TWENTY-THREE

The Manifold Power of Numbers: Lessons to Be Learned

RECENTLY, MY PROFESSIONAL reading activities have taken me to two subjects which intrigued me a lot. The main argument of the books and articles I read focused on the contention that there is power in numbers. Basically, power is derived from several sources known in physics, social and behavioral sciences as ***Critical Mass, Tipping Point, Spontaneous Reaction, and Leverage.*** Power is also obtained from biology and primate group behavior in the form of ***Swarm Intelligence and Influence.*** These concepts have great relevance to our corporations and not-for-business organizations.

How Does Critical Mass Produce Leverage

Let me briefly say a few things about how critical mass produces leverage. When a group of people is formed and that as time goes on the group becomes large enough to reach a tipping point. At that point it becomes a powerful force that it explodes (like in physics) into a spontaneous reaction. The group has reached a critical mass. That kind of power based on numbers can be used to leverage any position.

The same concept in social-psychological terms is applied to any number of human activities which have a pattern of slow development until an adequately sized group is formed. At this stage a tipping point is reached to tackle greater challenges facing the group domestically and globally. When an organization, such as an international organization, would sustain the critical mass required to propel progress – normally not available to small local organizations, it gains clout to be noticed and counted to make things happen for its constituencies with a clearly defined ideology spelling out unambiguously the values, vision and mission of the group.

The Application of Critical Mass Concept

The application of the Critical Mass concept is next. To illustrate what critical mass can do to even a small, politically insignificant group of people, let us look at the Inuit's case (aka Eskimos) when they wanted to be in control of their destiny. In the late 1970s, the Inuit people who had lived scattered throughout the vast regions of North America had to settle a territorial dispute with Canada.

A number of self-appointed leaders wrapped their collective brain around the notion of a self-governing nation. To their dismay, they soon learned that the task facing them could not be tackled by a handful of activists. If they work together, they would fight leviathans. Therefore, part of the calculus was the priority of coalescing. So, the leaders first met in 1977 to create "a world Inuit organization" to represent their people mainly in three countries consisting of Denmark (Greenland), U.S.A. (the State of Alaska), and Canada.

After having established a world organization, the Inuit next proceeded legally to claim lands from Canada. In 2005, the Inuit succeeded to build Jurisdictions called Nunavut (our land), a self-governing territory in the eastern and northern portions of the Northwest Territories. Nunavut is the largest and newest federal territory of Canada; it was separated officially from the Northwest Territories on April 1, 1999. Inuits' success is largely attributed to the formation of a worldwide organization which had reached a critical mass. Any such small nations can benefit from the super, manifold power generated by the critical mass phenomenon

The Power of Swarm Intelligence and Influence

The power of Swarm Intelligence and Influence is fast becoming known recently. We all know very well that humankind has learned a lot from the birds, the bees, the ants, and their likes. This type of knowledge has been anecdotal. Recently, however, we are reading empirically based study results. Very recent scientific studies have shown that those creatures (insects and other primates) produce a phenomenon known as **Swarm Intelligence,** which is not exhibited in its individual members.

If you were to look at an uninvited guest such as an ant on your kitchen counter, you would see that it is going frantically back and forth (like dancing a solo Cha Cha Cha) as though it is lost, it moves aimlessly in circles without any scheme (is not teleological). Biologists tell us that the ant is lost; the ant does not know what to do next. In a word, the ant alone is not "smart," even it is dumb contrary to what conventional wisdom has enshrined the insect as a very accomplished creature to be used as a model for human behavior.

However, when this "dumb" ant gets into his group, when it smells other ants and communicates with them, it becomes a crafty architect, an accomplished engineer; it is enabled to build bridges, pathways and even multiple-lane freeways; it is able to haul a delicious morsel of food hundredfold its weight. There is an uncanny power in numbers; sometimes it is called synergy (2+2=5). As a group, ants survive many catastrophes such as earthquakes, floods, wild fires, pesticide sprays and other chemical warfare waged against them.

The application of Swarm Intelligence and Influence is fascinating. The latest scientific discovery by some biologists in Africa is very interesting to see how mighty an individual becomes when in a group.

Elephants feed and often destroy acacia trees. Some biologists doing research in Africa have noticed that in certain areas where elephants were feeding on some acacia trees, some of the trees were untouched, unharmed despite their lush and bountiful vegetation. To satisfy their curiosity, they examined the unharmed trees for possible clues as to why the elephants had chosen not to use them as sources of food. They discovered that ants had taken residence in those trees.

Whenever the elephants approach the tree and spot the guardian ants lined up in front of their nest, they quietly and begrudgingly move over to the next tree instead of picking up a fight with these pesky guardian ants – which become formidable fighters when in groups. No smart elephant would like to have those ants in its trunk. Numbers have the magic power in them.

The Genius of Swam Intelligence Emulated

The genius of swarms is being emulated in human activities. While a single ant is not smart as an individual, but its colonies (i.e., group) has intelligence. The study of swarm intelligence is providing insights that can help humans manage complex systems, from truck routing to military robots. The comparison of swarm behavior to human interaction has become increasingly relevant. With the help of the computer, it is possible for human communities to behave like swarms of our own. In social networking communities the voice of one can quickly become the voice of one hundred or the voice of one million.

Swarm communities, how they form, interact, and disperse, and carry out certain tasks are irrevocably changing the landscape for humans to organize for efficiency and effectiveness. In swarm colonies there are no generals giving orders. Each and every member of the group acts to uphold the mission of the group without being told to do so. Such a behavior ushers in the concept of self-organizing, being self-motivated, being goal-directed, and working toward the good of the whole group.

Swarm Intelligence is now being also applied to corporate strategy in various areas of activities, such as organizing for maximum effectiveness, distribution and logistics, etc. Some companies such as Air Liquide have used swarm theory strategies to solve complex business problems such as routing trucks, channeling telephone calls across a network. To become more efficient, law firm managements are considering to become self-organized with a few "rules of thumb," rather than centrally controlled. In computer technology, swarming behavior resulted in a technique known as PSO, particle swarm optimization in which one has a number of options to choose from and one wants to know the best option in advance.

Wolves are unable to hunt big games such as buffaloes individually, but when they form a pack, the likelihood of success changes drastically. International organizations around the world are facing big challenges, too. By adoption of the critical mass concept and the swarm intelligence theory and practice would give them a better chance against their competitors.

CHAPTER TWENTY-FOUR

Post-Industrial Society and Its Speed of Recovery

L IKE MOST PEOPLE, since September 2008, I have been concerned about the recovery of the United States' economy out of personal and social reasons. I had borrowed money and thought I had made wise investments in real estate in a burgeoning economy. Then the sky fell. We had the freefall in stock prices.

My retirement nest egg in 403b in mutual funds was to crack from all sides. I knew that it was not a question of whether it will happen. It was a question of when. The clock was ticking and in less than four months' time I lost 37 percent of the value of my retirement funds. And the value of my stocks kept slipping by the hour! I faced many a sleepless nights.

Global Market Relative Recovery

As we all well know, the last decade was marred by the fear of another "Depression." Luckily enough, as we turned to a new decade, some financial pundits began to herald recovery around the world. Encouraged by the good news, I wanted to find out some evidence for the euphoria. My search brought me to a global market comparison for the relative recovery around the world prepared by the Charles Schwab Bank.

For example, the data showed loss in stocks from September 15, 2008 to March 9, 2009 to be -43.28 percent and the stock rebound gain to be 64.83 percent for the United States as a post-industrial economy. While, on the other hand, for Indonesia as a developing nation, the loss was -39.89 percent and the stock rebound was a whapping 161.92 percent.

I was immediately struck by the huge disparity in the relative recovery rate. Most post-industrial (based on services) economies' recovery rate was lower than their agricultural – and manufacturing-centered economy counterparts. Nearly forty years ago, Daniel Bell's prognostication rattled the smokestack landscape that economies based on service would outperform economies based on agriculture and manufacturing. I was confused and looked for a plausible explanation. I searched in behavioral science and not in economics.

Daniel Bell's Post-Industrial Society

Daniel Bell's 1973 book, ***The Coming of the Post-Industrial Society***, has created many controversies and spawned numerous debates over the years. Central to the idea of a post-industrial society is the forecast of contemporary social change due to the proliferation of information and knowledge ushered in by the invention of the personal computers.

Over two centuries later, however, the Industrial Revolution is still taking place in most parts of the globe. Many of the world's less developed countries continue to make the transition from a rural economy to an industrial one. On the other hand, economies of some nations with a long history in manufacturing have been moving on to become "post-industrial" as coined in the early 1970s by the American sociologist Daniel Bell.

According to Bell's theory, developed economies such as the United States are moving from agriculture and manufacturing to service industry. The focus has shifted from the blue-collar workers to white-collar workers, namely professionals, administrators, and to service providers in health, education, communications, and banking.

The most valuable stock of a professional is his or her possession of information. Therefore, equipped with a good knowledge of scientific research methodology and the aid of computers, the most essential economic activities are the gathering of data, their processing, and their interpretation by professionals such as accountants, market researchers, journalists, and other specialist consultants and advisors.

R&D The Path to Economic Progress

In practical terms, this emphasis on data acquisition and the derivation of information out of the data for decision making drives home the idea that businesses should allocate a higher priority to R&D (Research and Development). This type of change in the philosophical orientation is evidenced by the fact that the highest performing economies allocate hefty amounts of money on R&D.

I wondered again as I refreshed my memory of Daniel Bell's theory. How come "the highest performing economies" are slower in recovery than the other stodgy and slower economies? I could not find a satisfactory explanation other than my personal observation that in post-industrial economies, consumers could easily forego services while in an agricultural – and manufacturing-based economies consumers have to have necessities of life.

Self-Service Increases in Service Sector

For example, I used to have my car washed nearly every month when our (US) economy was humming before the inglorious September of 2008. Because of the sagging economy, I began to do myself certain service work that I used to commission others to do. Hence, the service sector suffered from unemployment when I shifted the service task to myself from others. This shift cannot be easily made in agricultural – and manufacturing-based economies. Consumers in developing economies have to have their basic needs taken care of by farmers to grow food and by manufacturers to process the food for consumption.

The gnawing fear of falling on hard times must have prevented many people to spend money on anything other than necessities. In my case, my retirement nest egg had shrank by 60 percent in eight months' time. I shuddered at the thought of my chicken hatching still born. Like millions of others, I had to increase the volume of do-it-yourself projects. Pruning trees, washing the cars, painting the house just to mention a few all came under my department to survive the debilitating loss of many years' investments in real estate, stocks, and mutual funds.

My fears have indirectly contributed to the demise of the economy as I refused to employ other individuals' services. I am even now toying with the idea of asking my soul mate to stay home and cultivate a vegetable garden in the backyard of the house and possibly begin to keep a menagerie of domestic animals for subsistence living! Thus the cumulative effect, perhaps gave rise to nearly 12 percent unemployment in California, the sixth largest economy of the world.

My explanation may sound insufficient to some of our readers; it is tantamount to fighting windmills. I have oversimplified the issue contrary to the methods used by economists who rationalize by observing past behavior to forecast the future without getting into the psyche of the consumer in a new financial environment with a new set of economic parameters.

I would like to invite scholars even Mr. Daniel Bell (born in 1909) to shed some light as to why it takes relatively longer for a post-industrial economy to recover than a developing nation's economy?

Prudence dictates that I should stop right here before I get further confused by dabbling in the science of economics which does not welcome behavioral scientists' meddling!

CHAPTER TWENTY-FIVE

Immigration Behavior: Toward a Social-Psychological Model for Research

Note: The author has collaborated with Ms. Zara Mokatsian on this chapter.

MIGRATION AND IMMIGRATION activities have been germane to both human and animal kingdoms. While migration is now largely undertaken by animals in a grander and more patterned approach, immigration has been specifically in the domain of Homo sapiens.

Migration and Immigration Contrasted

Animal migration represents a collective travel with long destinations. The act suggests premeditation and unwavering willfulness known to humans as inherited instincts. On the other hand, immigration has been a complex process beyond simple explanations. Contrary to animal migration, human immigration has been disorganized and sporadic.

Recently, biologists have identified five major characteristics that apply in varying degrees and combinations to all migrations. They involve prolonged travel that carries animals outside their familiar habitats. Their movements are rather linear (not zigzagging). They entail certain behaviors of preparation such as overeating for the long and arduous trek. They demand special allocation of energy. Finally, the migrating animals maintain a strong commitment to the greater mission, which keeps them undisturbed from side temptations and undeterred by challenges (e.g., storms) that would turn other animals aside when in non-migrating mode. The Long and perilous journey must be continued.

An example of their commitment for a course of action is the wildebeests' migration. Once the herd decides to migrate, none of the rivers such as the Massai River in Kenya, deeply infested by ferocious crocodiles, would deter them from crossing the dangerous, murky waters hiding death and destruction. One by one, they all plunge into the river toward greener pastures without any attempt to retreat to their old habitats. The exodus cannot be reversed regardless of any real or potential hindrance or hurdles.

Animals seem to migrate for two major reasons: one reason is for moving into a more favorable environment such as birds escarping harsh winters to warm locations; the second reason is to find abundance of good food for survival and propagation of their species. To a lesser degree, humans such as the Laps of Norway and the Mongols in Central Asia till today also migrate for the same two reasons as cited above: better climate and better food supply for their domestic animals.

Why Do Humans Immigrate?

When it comes to humans as to why they immigrate, economists and social-scientists provide us with a myriad of isolated reasons. An extensive review of the literature and meta-analyses of studies on immigration, leads one to assume that the main purpose of immigration can be subsumed under two major categories of incentives: material and non-material.

Material incentives are mainly economic benefits (e.g., better standard of living), while non-material incentives would consist of social-psychological reasons such as the chance for self-actualization. Studies on immigration, whether scientific or anecdotal, have isolated mainly material factors such as better opportunities, availability of jobs, better standard of living. Other researchers have proposed some non-material factors such as political strive at home, corruption of the government, no hope for self-actualization and so on.

Unfortunately, most of the explanations behind immigration have been limited to a very few factors. Meta analysis of these studies indicated that collectively there are many more salient factors contributing to a decision of an individual or family to immigrate into another country. These factors are needed to be considered as well in determining as to why individuals immigrate.

A Model of Immigration Behavior

For the sake of brevity, here is a multi-dimensional model of immigration behavior from a social-psychological perspective. Here are the major components of the Social-Psychological Model of Immigration Behavior as shown in Figure 1:

Figure 1

The Social-Psychological Model of Immigration Behavior

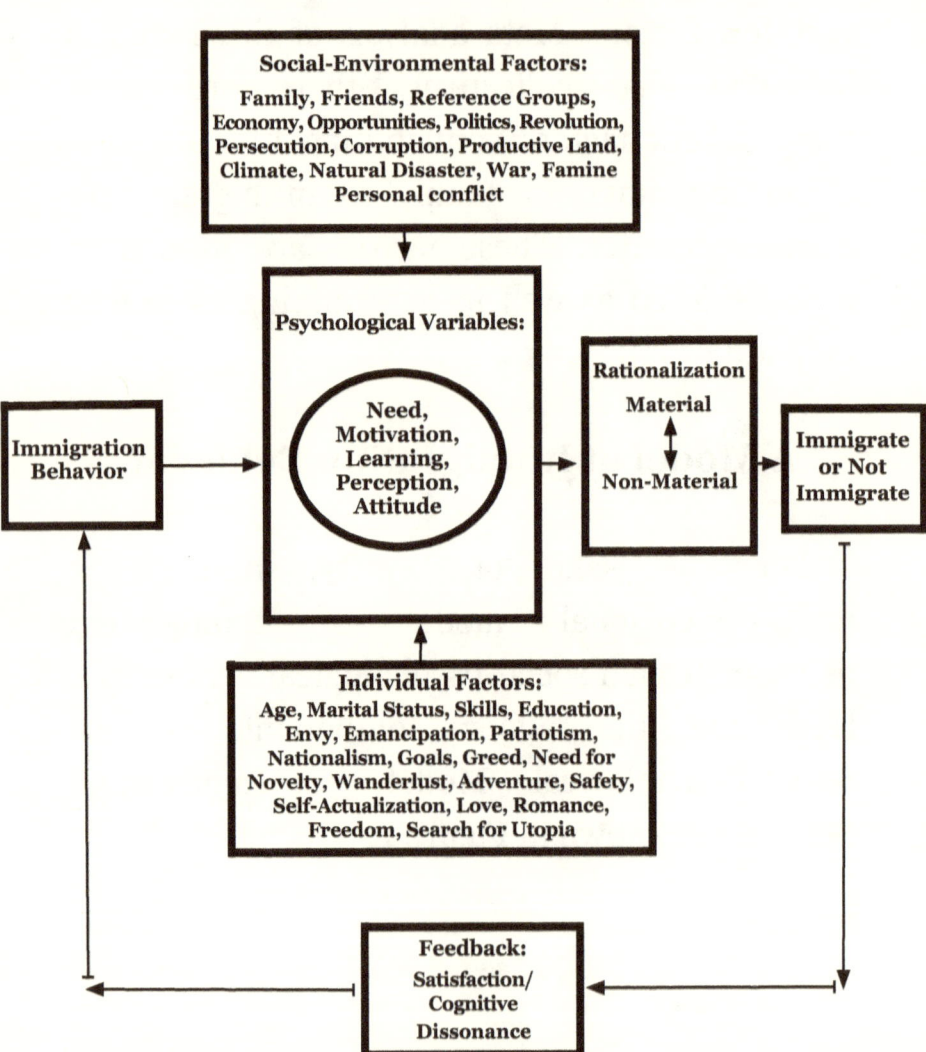

Social-Environmental Factors:

Family, Friends, Reference Groups, Economy, Opportunities, Politics, Revolution, Persecution, Corruption, Productive Land, Climate, Natural Disaster, War, Famine Personal conflict

Psychological Variables:

Need, Motivation, Learning, Perception, Attitude

Rationalization

Material

Non-Material

Immigration Behavior

Immigrate or Not Immigrate

Individual Factors:
Age, Marital Status, Skills, Education, Envy, Emancipation, Patriotism, Nationalism, Goals, Greed, Need for Novelty, Wanderlust, Adventure, Safety, Self-Actualization, Love, Romance, Freedom, Search for Utopia

Feedback:
Satisfaction/ Cognitive Dissonance

I. Immigration Behavior is to study the variables and factors contributing to the individual's decision process. The unit of analysis in the preceding proposed model is the Individual who is confronted with the challenge of deciding whether to immigrate into another country or to stay put in his or her country.

II. The Individual's psychological variables, such as Need, Motivation, Learning, Perception, and Attitude process the Individual's choice of Material or Non-Material incentives or a combination thereof as the Individual's primary reasons for immigration.

III. Social-Environmental Factors influencing the Individual's decision process consist of political situation, physical environment (e.g., climate at home country), economic conditions (e.g., job opportunities), etc.

IV. Personal factors which also influence the Individual's decision process are age, marital status, personal goals, etc.

V. The Individual's outcome of the decision process: immigrate to a perceived "utopia" or not to immigrate and stay at home.

VI.   Feedback is in terms of level of satisfaction and cognitive dissonance after the implementation of the decision.

The proposed Multi-Dimensional Social-Psychological Model of Immigration Behavior (to immigrate or not to immigrate) can be stated:

IB = f(P variables + S-E factors + Individual factors)

Where,

IB= Immigration Behavior

f= is a function of

P = Psychological variables

S-E= Social-Environmental factors

I= Individual factors

West Is Perceived as a Paradise

The Individual's Decision Process to immigrate or not to immigrate is a function of some of most of the above-mentioned variables and factors. To many young generations, the West is a "Paradise." Many students come to the USA to study with the expressed intention to return to their country upon graduation. After having enjoyed the opulence, the freedom, the availability of opportunities for advancement, usually they marry a local girl and stay. Then after a few years, they uproot their parents to come and live with them. Because of family ties, the parents decide to immigrate into a country not because of economic gains but because of their sons and daughters who had decided to remain in the host country.

Many scientists obtain government grants and are fascinated to study the epic migrations of animals requiring amazing endurance and boundless energy, yet many governments give lip service when it comes to the explanation as to why their people are leaving their ancestral lands to only settle in a foreign country.

The factors behind migration are easy to explain while immigration is complicated. The behavior of immigration is obvious to detect, it is objective, but why the individual immigrates is mixed with reality and wishful thinking. Even those who immigrate or plan to immigrate do not know for sure as to why.

The individual who has planned or who has already immigrated has to justify his or her act by putting the blame on, for instance, bad government or the monopoly of the oligarchs, etc. Some people have the tendency to invoke God or manifest destiny to rationalize their behavior. They do things in the name of God, Allah, Moses, etc. to ease their conscience in carrying out their inner desires for abandoning their motherland.

An Example from Social-Psychological Factors

Let us take "Persecution" from Social-Psychological Factors to see how it would influence the individual's immigration behavior. The Soviet Jews began to immigrate to Israel mainly because to avoid persecution. Hypothetically speaking, these Jews felt that they lacked freedom and considered themselves persecuted because of their religion. So, their decision to immigrate was based on non-material incentives or reasons (e.g., safety). Once in Israel, though, they experienced very low satisfaction from the move and even they had cognitive dissonance (of questioning whether they had done the right decision to leave the Soviet Union). The low satisfaction and the incidence of cognitive dissonance may have stemmed from the constant conflict with the Palestinian demands and their suicide bombers.

The Need for a Model for Research

We need a model, a theory to guide research. Europe, for example, has been concerned with the influx of immigration especially from the Muslim countries. It has been estimated that there are well over 20 million Muslim living in Western Europe alone, who defy assimilation.

Likewise, the Christians of the Middle East are immigrating to the West in droves, mainly due to the internal political unrest. the Christian population nowadays are leaving the Middle East not to seek greener economic pastures, but rather for safety from the dangers of revolutionary strive in the countries they happen to live.

The proposed model will put the beast of immigration on the table for scientific analysis, explanation and prediction. In other words, this model would serve as a source, a unified theory to generate hypotheses to be empirically tested and validated.

Policy decision makers could use the model in determining what percentage of the population is likely to immigrate and for what main reasons in order to device strategies to countervail these tendencies with the hope of retaining them at home. Therefore, this model is directed to my colleagues who have a lively interest to pursue the study of immigration in an organized and scientific way to advance our knowledge of immigration behavior which is becoming a more essential topic for developing nations which have small populations and yet they suffer from the loss of the most precious asset of the country through the brain drain phenomenon.

History has lessons to teach our new generations of policy decision makers: whenever a nation constantly loses its population to immigration to other countries, that nation is eventually doomed to oblivion. More bluntly stated, a nation without a vision will perish.

CHAPTER TWENTY-SIX

Nanotechnology: The Split Personality Symptoms

WHEN IT COMES to Nanotechnology, the euphoria of its proponents is louder than the resounding alarm of the doomsayers. Based on my extensive research on this promising science over the years, I have come to the conclusion that, like all other new sciences, debate rages over its positive and negative environmental and societal effects. It so seems that this new field has a dissociative disorder bordering on having a split personality.

The Split Personality of Nanotechnology

Split personality here is not meant to insinuate a neurosis in which the personality becomes dissociated into two or more distinct parts each of which becomes dominant and controls behavior from time to time to the exclusion of the other parts. Here it is meant that nanotechnology has multiple personality of the good, the bad and the ugly.

"The Good" Nanotechnology:

The positive aspects of nanotechnology are well documented. We only need briefly to cite a few of its major benefits. One big good comes from its efficiency and environmental friendliness. Furthermore, molecular manufacturing ensures that very little raw material is wasted.

The financial benefits for countries involved in nanotechnology are bright. Nanotechnology is expected to be a $2.6 trillion market by the year

2015. The race is now to get a large share of this lucrative market.

Numerous writers have pointed out the positive traits of nanotechnology by discussing its life-saving developments in medicine, overcoming the world's current environmental problems, in electronics, in robotics, etc. The list of its potential is impressive and long. Let us now present some of the "bad and Ugly aspects of this emerging field.

"The Bad and the Ugly" Nanotechnology:

Many scientists hail the positive benefits derived from nanotechnology, but only half as much discuss about its bad aspects. The positive aspects are readily accepted, but what about the bad ones? Is nanotechnology worth all the time and money spent on research and development like it was done for electricity and nuclear energy? Or are we dealing with another case of DDT and asbestos posing harm and destruction of the environment and society at large?

Before it is too late, we need to understand and be able to control the beast in nanotechnology before it is wholesale unleashed upon our world.

The Negative Side of Nanotechnology

Here are some negative consequences predicted for nanotechnology:

> **Health Hazard** – Nanoparticles have been shown to be absorbed in the livers of research animals and even cause brain damage in fish exposed to them after just 48 hours. If nanoparticles can be taken up by cells, they would be able to enter our food chain through bacteria and be deadly like mercury in fish, pesticides in fruit and vegetables or even hormones in meat.

> **Air Contamination** – The carbon nanotube which is twenty times stronger and lighter than steel resembles asbestos fiber. What would happen if they are released into the air we all breathe? Since they are carbon-based particles, they would not immediately show the negative

symptoms in our bodies. Thus, making them difficult to detect in a timely manner.

➢ **Environmental Pollution** – In reality, nanomaterials are as strong as diamonds, how decomposable would they be? The burning question is will they litter our environment further? We already have a major problem on our hands like nuclear waste or space litter? Will self-replicating nanobots (which are necessary to create the trillions of nanoassemblers needed to build any kind of product) spread as quickly as a new strain of virus? This is an ominous danger, to say the least.

➢ **Privacy Violations** – Based on the core characteristics of nanoscience, products would shrink in size; thus, giving rise to eavesdropping devices which become invisible to the naked eye and more mobile making it easier to violate our privacy. Tiny cameras would be excellent tools for spying. Soon James Bond will be using nanogadgets in the new films. These devices would be small enough to plant them into our bodies, mind-controlling nanodevices may be able to affect our thoughts by manipulating our brain processes.

➤ **Fostering Terrorism** – Terrorists are keen on having advanced military weapons, especially when they are small and portable to be used against us. Nanotech may create new forms of torture or simply disassembling a person at the molecular level. Radical groups could let loose nanodevices targeting to kill any ethnic person with a certain religion or skin color.

➤ **Weapons of War** – Because of nanoinstruments, concerns include the possibility of military applications of nanotechnology such as in implants and other means for solider enhancement which is currently being developed at the Institute for Solider Nanotechnologies at MIT. There is also the fear of the possibility of developing chemical weapons. Chemicals developed from nano particles will be more dangerous than the present chemical weapons.

➢ **Surveillance Problem** – Because of all the potential abuses of nanotechnology, many concerned experts advocate a viable system to regulate and monitor nanotechnology developments. Since nanotechnology laboratories can be very small and mobile, surveillance must be carried out everywhere. This is a herculean task beyond any feasibility of success.

➢ **Proliferation of Robots** – Since nanotechnology operates on tiny scale, the manufacturing of robots are made possible. Some fear that the robots will take over the world, especially those designed to be completely autonomous. As such, they are bound to be controlling the world one of these days.

➢ **Loss of Incentive to Trade** – It is expected that nations can make anything they want at little or no cost. The relevant question is will these nations lose all incentives to trade.

➢ **Fear of the Unknown** – The field of nanotechnology is a relatively new concept.

Its effects are lagged. Therefore, experts do not know its harmful side effects. Currently, more than twenty countries are producing with nanoscale materials. New nano products are entering the market at a rate of three to four products a week. Many of these products use chemicals that are not known to the consumers and may prove to be fatal.

> **False Hype** – many fear that nanotechnology will end up just like virtual reality. At one time, the confidence of its proponents was lofty until recently it fizzled. Now, the hype for nanotechnology is extremely high, what if we were to find out that we had been wrong about this new science, how can we reverse its harms when the market is inundated with new nanotech products every week?

Nanotechnology Helps and Harms

There is no question that nanotechnology has a split personality. It has the potential to help and also to harm. How can we weigh the bad vs. the good? Would it be fair to sacrifice human lives due to harmful nanoparticles in the environment just to have stain-resistant garments? This is a moral issue which should be addressed by those who are pursuing today the development of nanotechnology.

On the other hand, nanotechnology is a promising field, should our fears and concerns make us relegate it into oblivion? In 1876, Western Union dismissed the telephone as being useless as a serious means of communication. Because of the potential harms of nanotechnology, we should not dismiss it nor condemn it as being contrary to society's interests because of its bad and ugly aspects of its personality. However, we need to exercise extreme caution in its applications in order to be able to winnow the good from the bad and the ugly.

CHAPTER TWENTY-SEVEN

The Dynamics of Brain Drain during World Economic Decline

THE 2008-2012 GLOBAL Recession, also referred to as the Great Recession, is characterized as global economic decline that began in December of 2007 and nosedived in September of 2008. Now, it is has loomed as a major global recession caused by various systemic imbalances and sparked by the outbreak of the 2007-2012 global financial crises. As we all painfully know now, the global recession has affected the entire world economy by devastating some countries more than others.

The Nature of Brain Drain

Presently, the world economy is at its lowest ebb. Recovery has been very slow or nonexistent for some countries, thus widening the gap between the developed and developing nations. During an economic decline, the brain drain of developing nations is obviously accelerated. Brain drain is also known as "The Human Capital Flight". The process can be defined as the mass emigration of technically skilled people from one country to another country. The reasons behind the brain drain can run the gamut of political instability of a nation, lack of opportunities, unemployment, health risks, personal conflicts, joining of friends and relatives, and the establishment of new relations.

Brain drain is viewed as "human capital flight" because it resembles the case of capital flight in which mass migration of financial capital is involved. The government sponsors many costly programs to educate and train its youth only to see them immigrate to other countries after completing their programs.

The term brain drain was introduced by observing the emigration of the various technicians, educators, doctors and scientists, from various developing countries (including Europe) to more developed nations like USA. Now this phenomenon of brain drain has an opposite effect for a country into which people are immigrating. In this case, the brain-drain of a nation becomes the brain-gain for the host country.

Usually, all developing countries including China, India, Pakistan, Turkey, Iraq, Iran, Egypt, Armenia, and all the former Soviet republics are suffering from brain drain. Developed countries like USA, UK, France, Germany, etc. are having brain gain from this type of immigration.

The Dream of Western Paradise

The Western world, dubbed as Western Paradise, has always been the beneficiaries of brain drain. The "haves" have always had something attractive to the "have-nots." The opulence, the opportunities of the West has hijacked the brains of the youth of developing nations. For instance, many international students subsidized by their governments to obtain higher education in the West remain in their host countries after graduation. In this way, the home country becomes impotent in science and technology.

Brain drain has always been of major concern of many developing nations. Many articles and books have been written on the primarily negative effects of brain drain, but a very few nations have done something substantial to reverse the flow of skilled talent from going abroad.

Many attempts have been made by developing nations to reverse the flow of brain drain, but only a few could achieve some appreciable results. However, Egypt is one nation credited with a pioneering project to exploit the brain drain by turning it into a brain gain.

Egypt's Project for Brain Gain

In the heart of Cairo, a top-class research facility is currently attracting talented young Egyptians who might have gone elsewhere to learn and eventually stay there. Also, this research facility is serving as a magnet to students to return to Egypt after graduating from foreign universities.

Briefly, this pioneering project is taking place at Egypt's renowned National Research Center. This scholarly institution also houses the Nobel Project (Egypt's National Project for Scientific Renaissance) that provides young researchers who have left the country a chance to return to top-class facilities.

Acceptance to the National Research Center was initially based on candidates having a Ph.D. from foreign universities when populating ten newly renovated laboratories. Subsequently, installation of modern equipment at the center and the help in reducing red tape have lured back many promising "post doctorates" from Germany, Spain, the US, and the UK.

Egypt's success, no matter how modest, should be emulated by other developing nations who are losing talented young men more than ever before during this long-lasting world economic decline.

When a young graduate is retained by his or her home country, a young mind is gained. After all, the most precious asset of a country is not its size, its mines, its fields, its rivers, its mountains, but its quality of its people. As John Green (an American author) once said: "We all matter – maybe less than a lot but always more than none."

CHAPTER TWENTY-EIGHT

Gender Mainstreaming Goals: Shattering the Glass Ceiling Syndrome

T HROUGHOUT THE LONG history of humankind, women have been employed as a second fiddle to men. Deprived by lack of the oxygen of opportunities due to social imperatives and cultural exclusives, women, especially of poor countries, need internal and external assistance to stand on their own socioeconomic feet. Women around the world have been relegated in most areas of human endeavors despite the fact that they have been definitively found to be talented, tenacious, and energetic.

Gender Mainstreaming Paradigm

Gender mainstreaming has been proposed to create conditions to provide women equal opportunities to succeed socially and economically for personal as well as for national progress such as in the areas of eradicating hunger, improving literacy, preventing deadly diseases, etc. In other words, the intent has also been to put women in the path of empowerment.

Even in today's industrialized nations, only a handful of women have succeeded in bursting through the proverbial glass ceiling. Out of thousands of large U.S. corporations, less than half a dozen of women hold CEO positions in this highly industrialized nation.

Internally, some social entrepreneurs have interceded to improve the lot of women in poor rural areas such as Bangladesh's Muhammad

Yunus of the Grameen Bank who is awarded the 2006 Nobel Peace Prize for being one of the pioneers of micro-credit lending to provide seed money to women. Externally, the United Nations have stepped in to promulgate a strategy with eight important goals targeted to be achieved within less than ten years.

The United Nations Promise of Eight Goals

As citizens of the world, we must all participate in solving the problems of the poor of the developing nations. The United Nations Millennium Conference Summit meeting of 189 nations held in New York recently made a firm commitment to achieve the following eight goals set by the United Nations with a success rate of accomplishing fifty percent (50%) of each goal by 2015 as is enumerated in Table 1.

Table 1

United Nations Eight Goals to be 50 Percent Accomplished by the Year 2015

Goal 1: Eradicate Extreme Hunger and Poverty (Objective 1 – Between 1990 and 2015, reduce by half the proportion of people whose income is less than $1 a day);Objective 2 – Between 1990 and 2015, reduce by half the proportion of people who suffer from hunger)
Goal 2: Achieve Universal Primary Education (Objective 1 – Ensure that, by 2015, children everywhere, boys and girls alike, will be able to complete a full course of primary schooling)
Goal 3: Promote Gender Equality and Empower Women (Objective 1 – Eliminate gender disparity in primary and secondary education, preferably by 2005, and in all levels of education no later than 2015)
Goal 4: Reduce Child Mortality (Objective 1 – Reduce by two-thirds, between 1990 and 2015, the under age-five mortality rate)

Goal 5: Improve Maternal Health (Objective 1 – Reduce by three-quarters, between 1990 and 2015, the maternal mortality ratio)

Goal 6: Combat HIV/AIDS, Malaria (Objective 1 – By 2015, to reduce by half and begin to reverse the spread of HIV/AIDS; Objective 2 – By 2015, to reduce by half and begin to reverse the incidence of malaria and other major diseases)

Goal 7: Ensure Environmental Sustainability (Objective 1 – Integrate the principles of sustainable development into country policies and programs and reverse the loss of environmental resources; Objective 2 – By 2015, to reduce by half the proportion of people without sustainable access to safe drinking water and basic sanitation; Objective 3 – By 2020, to realize a significant improvement in the lives of at least 100 million slum dwellers)

Goal 8: Develop a Global Partnership for Development (Objective 1 – Develop, further, an open, rule-based, predictable, nondiscriminatory trading and financial system; including a commitment to good governance, development, and poverty reduction both nationally and internationally; Objective 2 – Address the special needs of the Least Developed Countries; including tariff-and quota-free access for the Least Developed Countries' exports, enhanced program of debt relief for heavily indebted poor countries [HIPCs] and cancellation of official bilateral debt, and more generous official development assistance for countries committed to poverty reduction; Objective 3 – Address the special needs of landlocked developing countries and small-island developing states (through the Program of Action for the Sustainable Development of Small Island Developing States and 22nd General Assembly provisions; Objective 4 – Deal comprehensively with the debt problems of developing countries through national and international measures in order to make debt sustainable in the long term)

Prioritization of Goals
Based on Perception

Considering the depth and the width of the prevailing poverty conditions in most developing countries, the United Nations' slate of lofty goals would be difficult, if not impossible, to accomplish within the given short period of years. Lacking prescience, policy decision makers will not know which goal to pursue first. To avoid succumbing under the overwhelming pressure of the many tasks and programs needed to implement the projects designed to accomplish the UN goals, therefore, a method of prioritization based on the perception of those involved in gender mainstreaming in their country would facilitate the order of goal selection process.

Given such a necessity of information, studies are needed to be conducted to determine the relative importance of the eight goals through the perception of the poor people in rural areas. Once the research instrument is constructed, data could be collected from the rural areas of any interested country which intends to implement programs in the hope of achieving the UN goals in the order of perceived importance. I wholeheartedly would like extend an invitation to our scholar contributors to engage in research on the prioritization of the U.N. goals and other relevant issues to alleviate the deplorably substandard living conditions in the world.

Against the backdrop of increasing globalization due to advances in communication and transportation, national concerns have become world challenges. For example, the U.S. is not only concerned about taking care of its needy, but also to participate in the eradication of hunger in the world. For want of sufficient resources, it is necessary to prioritize the goals which would identify the pressing projects needed to carry out in an attempt to achieve these goals. Therefore, lack of resources of the poor countries dictates the prioritization of the UN goals. Moreover, the need is acute to empower women of the world to stand shoulder to shoulder with men so as to combat world problems of hunger, disease, etc. and also to present opportunities for women to self-actualize through the fulfillment of their potential by shattering the artificial glass ceiling inequitably and inordinately imposed on their initiatives and talents for so many centuries.

CHAPTER TWENTY-NINE

The Return to Communitarianism: Saving the Sinking Ship

HUMANS ARE BORN in communities and they die in communities. Around 3000 B.C., humankind had to establish communities in the Neolithic period when a group of Homo sapiens (i.e., hunters and gatherers) had decided to settle down in one place to grow their own food. In the pantheon of human achievements, agriculture reigns supreme for it gave the essential infrastructure for building a civilization upon it.

We All Live in a Community

Till today, everyone, including those who refuse to take part in communal activities, lives in a community. In fact, many people belong to a multiplicity of communities defined by activity or mutual interest. One might join a social club, a church, a mosque, a synagogue, or parent-teacher association. All these organizations bring members together under a common system of purpose, rules and values to occupy the same territory. For example, the Journal American Academy of Business, Cambridge is a community of various individuals working as a team committed to producing a quality periodical to serve as an intellectual forum to advance our knowledge in business and other related areas.

Communities provide local level of social interaction which usually serves as the stepping stone between family and nation. Toward the end of the 20th century, many social commentators decried the passing of the community and many politicians began to promise a return to lost community values.

To Be Communitarian or Not to Be?

For ultra conservatives, though, any deviation form capitalism is tantamount to committing the cardinal sin. Rush Limbaugh (the radio-talk show host), for example, has denounced President Barak Obama's Stimulus Package policy program as being basically socialistic by publically declaring during a speech "I want Obama to fail!" (On account of the former's perceived deviation from capitalism and individual liberties).

Casting aspersions on President Obama sounds utterly unfair for most of U.S. presidents have had communitarian tendencies and programs. President Bill Clinton, for example, was open about his support for much of Amitai Etzioni's communitarian philosophy. Whether he applied it in his policy programs is questionable.

Likewise, President Bush during his 2000 presidential campaign was a "conservative communitarian" advocate even though he failed to implement his policies. Bush's cited policies included economic and rhetorical support for education, volunteerism, and community programs, as well as social emphasis on promoting families, character–building education, traditional values, and faith-based projects.

At times, Western civilization seems to have found values such as individualism, choice, and freedom to be in conflict with community life. In the 1980s, however, out of political and social concerns about the disintegration of social cohesion gave rise to a new social and intellectual movement called communitarianism.

Adherents of communitarianism, called liberals in the American sense or social democrats in the European sense, generally share the position on issues relating to the economy, such as the need for environmental protection, public education, but not on most cultural issues. It is interesting to note that communitarians and conservatives generally agree on cultural issues such as support for character education and faith-based programs. However, communitarians do not support the laissez-faire capitalism staunchly embraced by U.S. conservatives.

A Synopsis of Communitarian Philosophy

Communitarian thinking is not new nor is it an American invention. Its roots can be traced to ancient Greece and the Old and New Testaments. Modern-day communitarianism began in the Anglo-American academia as a reaction to John Rawls seminal book *A Theory of Justice*.

Drawing heavily upon the ideas of Aristotle and Hegel, some political philosophers such as Alasdair MacIntyre, Michael Sandel, Charles Taylor, and Michael Walzer disputed Rawls' assumption that the principal task of government is to secure and distribute fairly the liberties and economic resources individuals need to lead freely their chosen lives. Communities have equal, if not greater, responsibilities in fostering and nurturing values.

Central to the philosophy of communitarianism is the concept that people are shaped by the values and cultures of communities. More importantly, individual rights must be counterbalanced by social responsibilities. One of the proponents of this thesis is Amitai Etzioni whose aim was to redress a balance that is gyrating too far toward personal autonomy and away from community values. Doing things together reinforced individual's sense of social responsibility.

President John F. Kennedy, epitomized this philosophy of social responsibility by stating in his speech: "Ask not what your country can do for you, ask what you can do for your country." In the same vein, but less eloquently, President Obama has said on several occasions before and after winning the election, that we should all do our collective social responsibility and do our individual share for the good of the community and the nation.

Today, we are all experiencing exceptional times. The U.S. business and industry are caught in the vertigo of a recessionary spiral. Unfortunately, we have fallen on hard times. What can the government alone do? What can an individual do? What can a community do? To get us out of this metastasized economic crisis, we should never forget that many social goals require partnership between public and private groups.

This is not to propose that government should be replaced by communities, but that the government should empower communities with strategies of support in the form of revenue-sharing and technical assistance to establish the structures of civil society, public-private cooperation, especially provide positive rights to the people.

Positive Rights as the Core of Communitarianism

Positive rights are also at the core of communitarian philosophy. These rights include state subsidized education, state subsidized housing, a safe and clean environment, universal health care, and even the right to a job with the concomitant obligation of the government or individuals to provide one as the need arises. To this end, communitarians generally support social security programs, public works programs, and laws limiting such things as pollution. Collectively, these rights are forming a "social capital" which belongs to all citizenry.

The major objection leveled against communitarianism is that by providing such rights, they are violating the negative rights of the citizens, rights to not have something done for one's fellow man. For instance, taking one's money in the form of taxes to pay for such programs as described above deprives individuals of property.

If President Obama wins, America wins! Our President is not a socialist. On a continuum from communism to capitalism, his policies would fall into the category of social democracy. Social democracy is not anti-capitalism by any stretch of the imagination. The alternative to his programs would be to let the economy delve deeper into recession and become a full blown depression during these turbulent times.

Investing in communities is not "building bridges to nowhere." It creates new businesses, increases jobs, spurs consumer spending which, in turn, accelerates the wheels of production. It was individuals obsessed with greed who brought the U.S. economy on its knees not communities with values. We all need to put aside our political orientation for a while and work with cohesion to support our President. To save the sinking "ship" as one of the richest and greatest nations of the world, we need to build again communities, the essential fabric of society, to get us out of this present debilitating economic crisis.

Our ancestors are eating Kentucky Fried Chicken.

CHAPTER THIRTY

The Transgenerational Effects of U.S. Food on Global Markets

THE UNITED STATES' current affluent lifestyle is fundamentally based on a free enterprise system, emphasizing consumption. By any measure, the United States consumes a disproportionately high share of global resources. While making up five percent of the world's population, it is the alpha user of virtually all traded commodities such as corn, copper, and rubber including nearly 25 percent of the world's energy. The U.S. addiction to oil is well documented and it is caught in an inflationary spiral.

The Drive for Overconsumption

As a result of the drive for overconsumption, the marketplace is bombarded with an ongoing proliferation of new products and services. Fierce price competition is relentlessly forcing corporations to come up with cheaper and artificial substitutes to be used in their products. Due to expansion in globalization, much of our food products have been exported to different countries around the globe.

Unfortunately, most of our food products are of junk food category. This problem is rather acute in the U. S. However, our concern should not only be about our own uneducated minority communities here in the States, but also about the people of developing countries who may be adversely affected because of U.S. corporations' practices of selling food products in overseas markets, which have no, or little, nutritional values and are the underlying causes of various diseases.

The problem has become global in scope: over one billion adults are overweight and more than 300 million of them are dangerously obese. Of this total, 115 million live in developing countries which had no similar problems just two generations ago. Obesity is associated with many dreadful diseases such as Type II diabetes, cardiovascular disease, hypertension, stroke, sleep apnea, and different types of cancer. Since developing nations lack adequate resources to treat these types of diseases, many people would die suffering to the end of their lives.

The Emerging Field of Epigenetics

The alarming situation is that the U.S. corporations are not only affecting the present people of those countries, but also extending the impact on their future generations. Recent burgeoning field of research has shown that our fathers' and grandparents' experiences affect later generations such as on sons and grandsons. This fascinating and emerging field is called epigenetics.

As you may already know, in biology, the term epigenetics refers to changes in gene expression (epigenes in Greek means born on or after). These changes may remain through cell divisions for the remainder of the cell's life. Sometimes the changes last for multiple generations. However, there is no change in the underlying DNA sequence of the organism; instead, environmental factors cause the organism's genes to behave or "express themselves" differently.

While a person's genome is inherited, his or her epigenes are shaped or formed by the environment as well. Thus, according to epigenetics, one's ancestor's diet, smoking habits, exposure to pollutants and levels of obesity could be affecting their descendants today. In turn, one's present lifestyle could affect one's children and grandchildren.

For example, identical twins inherit the same genes and look physically exactly alike, yet they may behave differently because of the environment as well and because of the lives of their grandparents (i.e., the epigenes). The fathers and grandparents air they breathed, the food they consumed, and the stresses they had experienced in their lives are passed on to their male offspring in the form of epigenes. Research-based evidence suggests that people who live unhealthily today could be stacking up problems for future generations and can directly affect one's wellbeing decades later, despite one has never directly experienced these things before on personal basis.

Transgenerational Effects Are Hereditary

Transgenerational effects have been seen as being passed down the female line before, but that these effects also operate on the male line has been pioneered recently by Marcus Pembrey (University College London) and his Swedish colleague Lars Olov Bygren (The University of Umea), who published their work in the European Journal of Human Genetics.

Recently, United Kingdom has blamed the increasing obesity of its population on American fast food restaurants such as McDonalds. When it comes to developing countries, most people are not well educated to be discerning and discriminating consumers to avoid American junk food. Therefore, they suffer the most. I would like to extend an invitation to our scholars to engage in research on the curtailment or elimination of unhealthy food and other relevant issues to alleviate the deplorably substandard nutritional properties of U.S. products promoted in the world markets.

As we all agree, the principles of justice, fairness, and social responsibility should also be included in the standard ethical practices of global corporations to ensure that their influences are both politically correct and ethically beyond reproach, possessing no adverse transgenerational effects on their host countries from selling junk food.

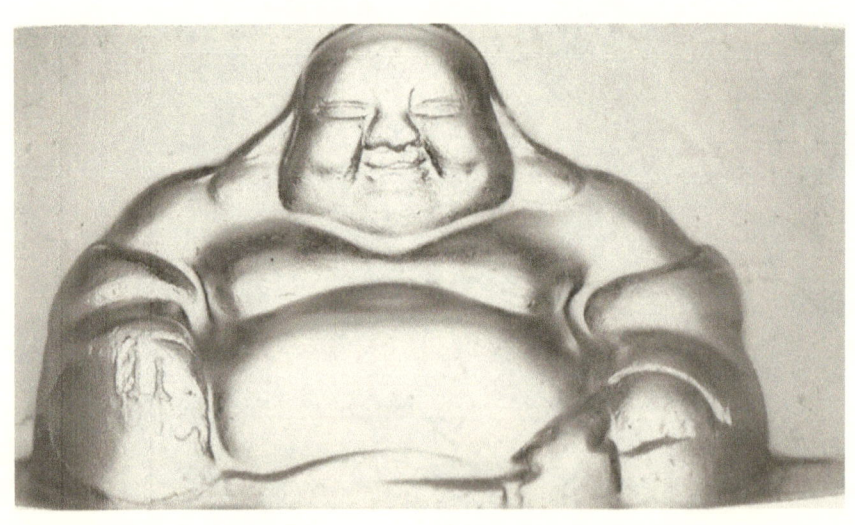

CHAPTER THIRTY-ONE

The Obesity Pandemic in the U.S.: A National Crisis

DUE TO THE alarming prevalence of obesity in the United States, many observers of the scene have unequivocally stated that the "disease" has reached pandemic proportions, affecting all ethnic and age groups. The crisis is beyond any doubt.

Prevalence of Obesity in the United States

According to a recent World Health Organization report (WHO, Global InfoBase, 2011), on the prevalence of obesity for adults, aged 15 and older, the United States stands clearly at the higher end of the continuum: 80.5 percent of males in the United States and 76.7 percent of females are overweight or obese (Body Mass Index of 25 or greater) as compared respectively to 3.5 percent and 6.3 percent in Eritrea (among the lowest percentages) and 96.9 percent and 93.0 percent in the Republic of Nauru (among the highest percentages).

Perhaps the most disturbing aspect of this phenomenon is that obesity rates appear to be on the rise. Figure 1 traces the rate of increase in adult obesity between 1960 and 2010.

Figure 1

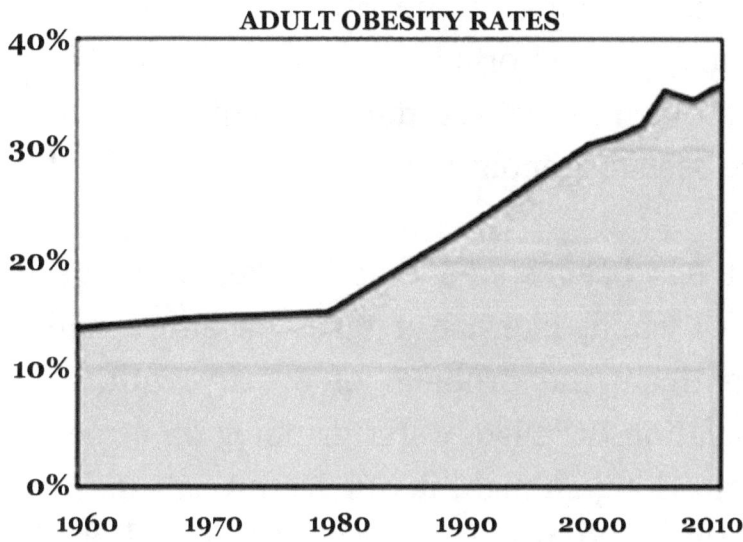

For example, between 1986 and 2000, severe obesity quadrupled, while extreme obesity increased by a factor of five, and similar figures were observed in children and teenagers.

Sadly enough, the human and economic cost of obesity is high, amounting to hundreds of thousands of deaths per year in the United States, and costing society well over $100 billion yearly, which exceeds the health-care costs associated with smoking or excessive drinking. While there are multiple factors explaining obesity, it appears that eating habits and lack of physical activity may be the most important culprits. Starting at the very beginning of this century, some efforts have been made to address the problem. However, the results of these efforts have been disappointing.

It has been stated that it takes a village to raise a child, likewise it takes an entire nation to mitigate the problem of obesity within the population. It is like water dripping on a rock; if it drips long enough, it will go through it. Thus, the United States cannot afford to stand on the sidelines while obesity is becoming rampant.

The Bilateral Framework to Mitigate Obesity

The coalition of all who are concerned about the crisis is long-overdue. A bilateral framework establishes a mechanism for cooperation between two groups of change agents for dealing more effectively with obesity as shown in Figure 2.

Figure 2

The Bilateral Framework
of Change Agents

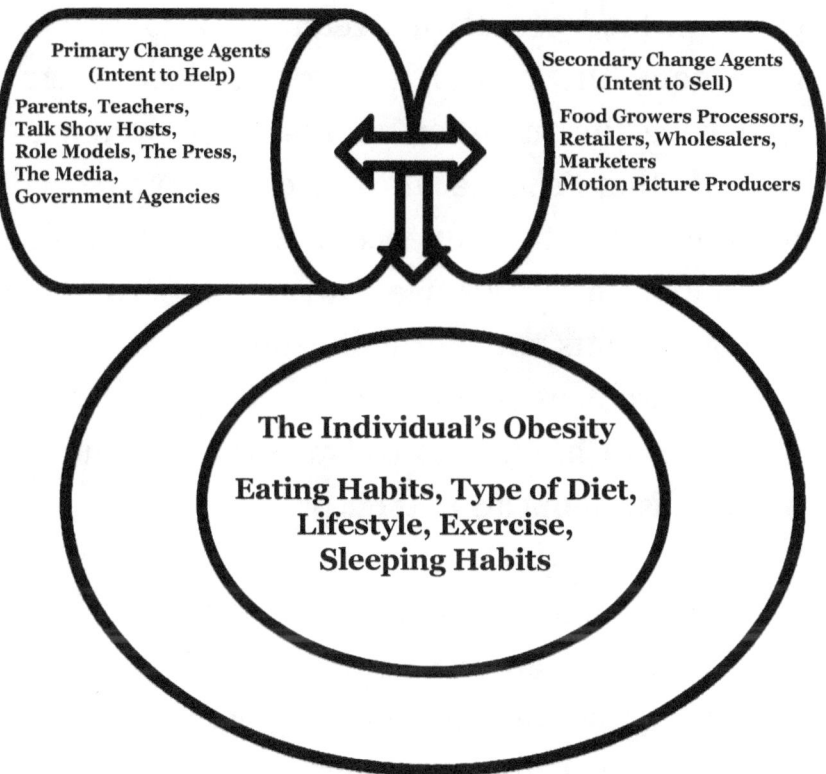

The bilateral framework consists of two sets of change agents: Primary Change Agents (e.g., parents, teachers, role models, the press, the media, talk show hosts, the government, etc. who have the intention to purely help the person) to

354 | CHALLENGES AND OPPORTUNITIES IN EXPONENTIAL TIMES

influence the obesity of the individual. These agents engage in more aggressive and sustained approach to tackle this problem.

Secondary Change Agents (e.g., food growers, food processors, retailers, wholesalers, marketers, the motion picture producers, etc. who have the intention to sell their products) can, if they choose to exercise their social responsibility, to influence the obesity of the individual in a positive way. The Secondary Change Agents are normally the source of the obesity problem in the nation. They also pose many health hazards to consumers by including dangerous chemicals in their products.

The Secondary Agents are driven by the profit motive and in order to survive and grow in a fiercely competitive market, they use ingredients that are cheap and the unwary average person would be ignorant of their damaging presence in the products they buy. For example, donuts could be prepared with honey or with a low glycemic sweeteners such as organic agave syrup, but instead they are dipped in a sea of sugar which is cheaper than honey. As nutritionists claim donuts are one of the cardinal culprits in contributing toward an individual's state of obesity.

The Miracle Teflon Coated Cookware

Another example to show the callousness of corporations in regards to the health of their consumers, let me cite the saga of the ubiquitous Teflon patented by DuPont. Teflon is a registered brand name for a man-made chemical known as polytetrafluoroethylene (PTFE). This product has been in commercial use since the 1940s.

DuPont made a killing out of this product since it has a wide variety of applications – because it is extremely stable, namely it does not react with other chemicals, and it can provide an almost frictionless surface. It is the dream of most cooks and housewives. Most people are familiar with it as a non-stick coating surface for pots and pans and other cookware. Additionally, it is used in many other products, such as fabric protectors.

Teflon also contains perfluorooctanoic acid (PFOA), also known as C8; it is another man-made chemical. This chemical is used in the process of making Teflon and similar chemicals known as flurotelomers. Although this chemical is burned off during the process and is not present in significant amounts in the final products, over time little by little it accumulates in the human and animal systems exposed to it.

Riding on the wings of research, the anecdotally-based knowledge about Teflon has been replaced by scientific testing to the deepest dismay of its inventor company. For years, Teflon coated cookware such as pots and pans were marketed as the non-stick miracle in cooking. However, the trend of recent research has shown the darker side of Teflon. Teflon, when hot, emits PFOA and PTFE. While Teflon itself is not yet a suspect to cause cancer, PFOA may be more of a health concern because it can stay in the environment and in the human body for long periods of time. Laboratory studies of rodents have found exposure to PFOA increases the risk of certain tumors of the liver, testicles, mammary glands, and pancreas in the experimental animals. The results of these studies are contested to have not provided sufficient evidence for causing cancer in animals; therefore, there is very meager evidence that it causes cancer in humans. Once asbestos was claimed not to have had the chemicals that cause any health hazard. Now, it is a common knowledge that asbestos causes cancer. Because of that many Western countries have banned its uses with which humans can come in contact.

The Major Culprits in Obesity

The obesity condition of the individual is caused by many factors. Paramount among these factors are: eating habits, type of diet, lifestyle, exercise and sleeping habits are some of the villains. It should be noted that these factors are in the adult individual's control. There is no denying the fact that there is a laundry list of reasons as to why someone is or has become obese. The physical or biological aspects of obesity are beyond the scope of our discussion here. Anyone who is navigating the road of obesity most probably recognizes obesity as a result of over consumption under exercising. The conventional medical thought is that eating the right food is not sufficient. Exercising is imperative. As you know, those who eat a lot, follow an unhealthy diet, lead a sedentary lifestyle, sleep few hours, and fail to exercise would give themselves the passport to obesity.

The Power of Primary and Secondary Change Agents

The Primary Change Agents with their explicit and implicit intentions to help and protect are able to influence the individual positively in his or her eating habits, type of diet, lifestyle, and exercise. For instance, conscientious parents would be concerned about junk food which would pay later negative dividends in their children's adult lives. Therefore, they avoid buying nutritionally suspect food products for the family.

The Secondary Change Agents have also the power to influence the individual positively in his or her eating habits, lifestyle, etc. However, such a corporate responsibility would go counter the bottom line pursuit. Therefore, the Secondary Change Agents simply give lip serve with tokenism when it comes actually remedying the obesity issue in this country. For example, the clothing industry should abstain from glamorizing obesity by promoting that "Big is Beautiful," "Full Figure Is Sexy," "One Size Fits All" instead they should demarket such items that encourage people to remain obese. While we need to clothe the obese as well and not let them go around like Adam and Eve in the Garden of Eden, but we should not enthusiastically push these products.

Clothing industry through fashion shows, can demarket clothing for obese people by simply excluding them. It sounds unfair, it sounds even discriminatory. But how could we reconcile with ourselves to let someone suffer from cancer due to being obese?

Now, this sounds like being really cruel to the individual, to society, and to the nation as a whole to let someone suffer from obesity.

Both the Primary and Secondary Change Agents are in position to influence the individual conquer obesity and get on the road to a healthier weight. To tackle the obesity problem, the Primary and Secondary Change Agents should take drastic measures before and after the incidence of obesity. The approaches are preventative, curative, and palliative. Of the three approaches preventative would be the best, yet it is least used by both the Primary and Secondary Change Agents. Instead, billions of dollars are expended on trying to remedy obesity since it is hazardous to one's health.

It should be noted that some of the Primary Change Agents such as parents, teachers, and role models, etc. are able to use the grassroots approach in nipping the obesity of the child in the bud. Parents should refuse to buy, for instance, sugar-coated breakfast cereal for their children. They should be armor-plated against junk food. This will be the ideal way since it does not involve difficult, expensive, and time-consuming corrective methods after the individual becomes obese in his or her more advanced age.

Secondary Change Agents are mostly guilty of contributing to the obesity of the young in tender age. For example, marketing uses many promotional strategies and tactics to entice children make their parents buy certain products. Including toys in breakfast cereal box is one example of unfair marketing maneuvering. Company spokespersons aver, tongue in cheek, that the primary intention for the toys is to make the child eat breakfast. The intention to sell and to compete fiercely against other manufacturers of breakfast cereal is the main reason for promoting products through giving out free toys.

The Supreme Ruler

Unfortunately, parents in the United States succumb to the demands of their children despite the fact that they are one of the earliest change agents in the lives of children. A case in point, in the 1980s when Captain Crunch, one of the leading brands of cereal, entered the UK market, to its management's huge surprise, the product was not well received. Upon examination of the issue through marketing research, it was found that the inclusion of toys in the box had alienated the UK mother for circumventing their prerogative to choose the cereal they wanted for their children. The UK mothers considered the tactic unfair and unprofessional. While the UK mother has control over what cereal to buy, the United State mother is virtually helpless. Children in the States rule. This is a well established fact.

Parents in the U.S., as being one of the Primary Change Agents, should use tough love in rejecting products that are not healthy for their kids. Kowtowing to the demands of their children runs the risk of turning these kids obese in their younger and older ages. The axiom "spare the rod and spoil the child" was used ages ago to be consonant with many faiths and religions that "those who spare the rod hate their children, but those who love them are diligent to discipline them." Sadly enough, parents in the States play the popularity contest without casting an eye on the future consequences of their actions.

The bilateral framework mentioned above covers the tip of the iceberg, but it is better to navigate in uncharted waters with a guiding compass rather than drift in the dark. There are many more variables excluded from incorporating in the framework such as the individual's genome, epigenes, the functioning of his or her hypothalamus, the transgenerational effects inherited from ancestors, just to mention a few.

The bilateral framework is designed to facilitate common efforts to develop strategies for combating obesity. The framework will also build upon cooperation between the Primary and Secondary Change Agents on a national scope, promoting efforts to reduce the incidence of obesity.

The Bilateral Framework Guides Research

Despite its limitations, the bilateral framework guides research systematically for it represents the problem in coherent picture. Furthermore, this model serves as a logical framework to build knowledge in the Primary and Secondary Change Agent variables. By simply testing hypothesis without an underpinning paradigm would not accumulate categorized knowledge in a given science just like a rolling stone would not gather any moss.

The United State society has become a salad bowl of various ethnic groups. When immigrants come here, especially from South America, they are dazzled with the abundance of food within the reach of the average family while in their native countries they had been deprived even of staple food. Once in the cornucopia of the richest country in the world, they throw themselves to buying and eating food with abandon. This kind of eating with vengeance, morphs them into heavy round obese shapes. The consequence of the frenzy of overeating makes them subject to many health conditions and hazards.

Various government agencies are also among the Primary Change Agents. While the Food and Drug Administration (FDA) is a force to contend with in the United States, their sphere of influence is limited for being understaffed and underfunded. French legislators are taking bold steps to control the widespread of fast food (aka junk food) in their country. Realizing the harm involved in fast food, legislators succeeded in demanding restaurants to state on their menus dishes that are homemade from fresh ingredients as "fait a maison." The idea is to promote traditional French dishes vs. fast food from frozen ingredients in that country steeped in culture. La cuisine Francaise is the sacred cow of the French. In this way, the customer would know which food is healthy and what to order knowingly.

FDA should also insist on the use of healthy ingredients even though we operate in a free enterprise system which is costing us billions of dollars in combating deadly diseases caused by obesity. For example, FDA could require companies to use organic blue agave or honey instead of sugar for sweetening the cereal breakfast which is extensively used for children.

In The Final Analysis

It has been stated that it takes a village to raise a child; likewise, it takes an entire nation to mitigate the problem of obesity within the population. It is like water dripping on a rock; if it drips long enough, it will go through it. Thus, United States cannot afford to stand on the sidelines while obesity is becoming rampant. Tough legislation may be just what the nation needs to start down the path toward reversing the increasing trend of obesity.

The coalition of the Primary and Secondary Change Agents working together is bound to reduce the incidence of obesity in the male and female population of the United States. In so doing, billions of dollars would be saved, many would not suffer from illnesses caused by obesity, and the country would end up having a healthy, vibrant population ready to contribute positively toward achieving a higher stratosphere of success.

On a lighter note, let me say that obesity is a disease. It makes everything taste good except salad. Let us enjoy our salad days, rather than our hospital years for untold complications gifted to us generously by obesity.

Part IV

Environmental Issues of World Concern

CHAPTER THIRTY-TWO

The Effects of Global Warming on Plants, Animals, and the Ecosystem

O F ALL THE major cities of the world, I enjoy revisiting London whenever the opportunity presents itself. Recently, I came across a photograph by certain Mr. Dean Whyte with the caption: London 2023: City under Water. The implication was that due to global warming London would be submerged under water by the year 2023.

The same photo was depicting the royal place in water with two mounted horsemen riding over it. Somehow the picture sparked my imagination to search further the apocalypse of climate change portentously signifying the advent of a far-flung calamity surpassing that of the Black Death. Mostly what I found in the literature was asserting not only the pending but also the manifestations of the doomsday.

Global Warming: A Hotly Debated Topic

To my ease of mind, I found out that the question of global warming is still being hotly debated. Those who believe in global warming phenomenon, however, outnumber those who think the whole thing is a hoax to sell "green" products – another marketing ploy par excellence. The proponents maintain that there is little doubt left that climate change is real, and global warming effects are taking place around the planet and are fast becoming all too obvious to ignore.

It would be easier to shelf the warning as a "hoax" and bury our heads in the sand. In my view, it would be better to error on the safe side and to begin taking measures to protect plants, animals, and the ecosystems from the eminent changes caused by the possibility of climate change.

To follow the old adage that says "It is better to be safe than sorry," it would be wise to consider the consequences of the possible effects of the events. The preponderance of opinion, scientific and lay alike, that global warming effects are leaving their mark on the planet as a whole, as well as on every plant, animal and ecosystem in some way or another.

Global Warming on the Environment

To avoid sheer speculation, let us discuss the most obvious effects of global warming on our environment. Due to space limitation, let me summarize what I have so far found about this nightmarish forthcoming coming event:

For global warming effects on climate, the global climate is likely to undergo the greatest increase in the temperature to occur over the polar region of Northern Hemisphere due to melting of sea ice and associated reduction in surface albedo (the light reflected by planet). There will be greater warming over lands than over the oceans. This would sound like music to the ears of the Inuit of Nunavut territory near the North Pole in Canada.

Changes in climate patterns would usher in floods, droughts, heat waves, extreme winter cold and snow fall, tornadoes, extreme storms, tropical cyclones/hurricanes/typhoons. There seems to be a lot of scientific evidence to support the statement that global warming is a major cause at work.

The global warming effect on the sea level predicts further increases are expected due to the melting of ice cover including Greenland and Antarctica. Complete melting of Greenland ice sheet would be caused by only an additional two degrees centigrade and would cause global sea level to rise by five to six meters.

The event would submerge a substantial number of islands and lowland regions. Among other regions, U.S. Gulf Coast and Eastern Seaboard (including the lower third of Florida), much of the Netherlands and Belgium, heavily populated tropical areas like Bangladesh, including world's major cities such as Tokyo, New York, Mumbai, Shanghai, Dhaka will become the Venice of the future (covered by water). If the complete loss of the West Antarctic ice sheet were to occur, this would lead to a 10.5 meter in the global sea level. Fortunately, this event might take several centuries to occur; however, it is possible that the rate of loss might be a lot faster.

Nothing is Totally Bad or Totally Good

Invoking the sociological concept of the Universal Functionalism (that nothing is totally bad and nothing is totally good), we would see that those countries that are geographically situated on high plateaus such as Iran, Turkey,

Armenia, Afghanistan, etc. would not be affected by changes in sea level. In fact, they would benefit from global warming in terms of increasing their growing season. While, on the other hand, those countries situated on lowlands, would suffer the consequences of changes in sea level.

There is a real apprehension among some small island inhabitants that their lands may totally disappear under water because of rising sea levels caused by global warming. For instance, the government of the Maldives (an archipelago of almost 1,200 coral islands located in the Indian Ocean with most islands lying just 1.5 meters above the sea level) is contemplating the purchase of land elsewhere in the world for the relocation of its nation. The fear stems from the fact that these islands will be totally flooded by rising sea levels. In addition to the loss of coastal areas, virtually every type of other ecosystems will be affected in some way or another.

The Sign of Things to Come

As a sign of what is coming, the tropical nation of the 33-island republic of Kiribati in the Pacific

Ocean, with its 100,000 inhabitants, is vanishing due to rising seas. The country is fast becoming uninhabitable. The sad part of the story is the question of where will the inhabitants go?

As for the global warming effects on the physical environment go, it would be disastrous to many terrestrial and marine ecosystems and habitats around the planet may very well be damaged and even disappear altogether. The ecosystems of coastal wetlands, salt marshes, and many mangrove swamps are most vulnerable ones. Rising sea levels as well as warming-induced catastrophes may affect millions of human settlements inhabiting them as well.

The UN Environmental Program has reported that 40 percent of the world's population lives in coastal areas less than 60 km from the shore. These populations and their environmental support systems are undoubtedly the most vulnerable groups at risk of the consequences in climate change.

Global warming effects on terrestrial ecosystem and habitats pose great dangers on forests and mountains. For instance, warmer climate encourages the growth of pests which destroy forests in unprecedented numbers. Pine beetle infestation of forests in British Columbia, Canada, which would have killed fifty percent of the pines within a few years of attack. Additionally, forest ecosystem also faces the great risk of wildfires due to the warmer climate. Science is establishing a sound relationship between global warming and wildfires which contribute immensely to the greenhouse effect. Amazon, humanity's lungs, is the biggest remaining tropical rainforest on the planet. By mid-century, increases in temperature and associated deceases in soil water are projected to lead to gradual replacement of tropical forests by savanna in eastern Amazonia.

Global Warming Effects on Mountains

The global warming effects have also been noted on mountains. Mountains cover roughly 25 percent of the Earth's surface and provide habitat of plants, animals, and people inhabiting these areas. Apart from an increased number and intensity of forest fires and reduced diversity of wildlife, another significant effect that climate change will force upon the mountainous ecosystems is the eventual melting of their snow cover and retreat and disappearance of glaciers.

It has been reported that rising global temperatures on rich biodiversity of tropical rainforests and other animal ecosystems are at an extremely high risk of disappearance. At danger of extinction are anywhere from 20 to 30 percent of species are highly at risk of extinction should the average warming exceed 1.5 to 2.5 degrees centigrade. If the temperature increase were to exceed 3.5 degrees centigrade, projections of extinction would rise to 40 to 70 percent of species around the globe.

Many Seasonal Processes Are Also Affected

Scientists have begun to witness many seasonal processes also affected by global warming. Here is a short list of the changes: earlier leaf production by trees (I personally noticed that on the trees in my orchard), earlier greening of vegetation, changed timing of egg-laying and hatching, changes in migration patterns of birds, fish and other animals, and reductions and re-distributions on populations of algae and plankton which threatens the existence of fish and other animals that heavily rely on algae and plankton for stable food.

There is also very damaging effects on marine ecosystems and habitats due to global warming. For example ocean acidification has been established beyond any doubt. Our oceans are the world's biggest carbon sinks and have absorbed about half of all anthropogenic carbon dioxide since around the 1800s. However, such a vital environmental service comes at a big price: the oceans have become significantly more acidic as a result of this process. Ocean acidification has been held responsible for damaging some of the most beautiful creatures on the planet such as coral reefs as well as other shell-forming organisms. Animals on the list of eminent extinction are the tropical frogs, amphibians, and the orangutans. The list of animals at risk of climate change will, naturally, become longer and longer as the planet gets hotter and hotter.

Global Warming Effects on Homo Sapiens

Finally, let us briefly discuss the global warming effects on the Homo sapiens as the top bananas at the food chain and as the prime culprits of movers and changers of the physical appearance of the Earth.

Since virtually all types of plants, animals and environmental ecosystems will be affected by global warming, the human society and all its complexity cannot escape the nondiscriminatory, unpredictable, wide-ranging and far-reaching tentacles of global warming consequences. In view of the most important areas from the standpoint of well-being, humans will be affected in terms of shortage of water supplies, food supplies, and human health.

As we all know, water makes up 70 percent of the human body's chemical composition. It is, therefore, an essential resource without which we cannot exist. This precious resource is now under serious threat from global warming. By the middle of the 21st century, water availability is projected to temporarily increase in higher latitude and in certain wet tropical regions mainly due to melting of mountain glaciers, and decrease in drier parts of the tropics and subtropics, especially in summer months.

Contrary to expectations because of melting of ice, the dry areas of the world will get even drier and will suffer severe droughts, especially southern Africa, Middle East, Western North America, and Western Australia. Energy generation will also be affected by water scarcity. Reduced water supplies will have negative effect on the power plants which depend on water for their functioning such as dams, nuclear plants, flour mills, etc.

Agricultural productivity for food supplies in general is expected to decrease as the temperatures rise. Agricultural food production will be further affected by extreme weather events such as droughts and floods. The warming of oceans will have a negative impact on commercial fisheries. Famine will increase around the globe due to shortage of food.

The water quality will also be negatively affected by heavy precipitation events which may contaminate drinking water supplies. Stagnant water pools will harbor all sorts of virus harboring insects which eventually weaken the health of the human system around the globe through the globalization process where if a virus emerges it in no time it spreads to the rest of the world due to the advances in transportation. Human health is under great stress from the climate change. We have already seen an example of the damage caused by a heat-wave that hit Europe in 2003, claiming 35, 000 deaths. Floods, tropical cyclones, and other extreme events will be another major cause of human deaths.

Infectious Diseases to Rise

Of all the debilitating effects of global warming, is the expectation of infectious diseases to rise. It is a well-known fact that disease-carrying insects breed prodigiously in wet hot conditions, especially in tropical countries. Millions of people will suffer from malnutrition on account of shortages in food supply. This, in turn, will lead to weakened immune systems and general health deterioration of humans.

In sum, the human condition will deteriorate with the global warming effects. To survive under austere conditions, humans will resort to violence such as killing, war, and even genocide. Many people are not aware of the fact that it was global warming effects that lie at the very heart of the well-known, little understood, tragic ethnic conflict in Darfur, Sudan. The conflict that began in 2003 was preceded by decades of drought, desertification, and overpopulation in this African nation. This sad situation forced camel-herding nomads from the Arab Baggara tribes searching for water for their livestock to travel south, which was heavily populated by non-Arab farming communities. The fight ensued for scarce resources the country could provide its people. Due to racial differences, the south had to be cleansed of the indigenous tribes so that the nomadic Arabs could survive.

This shows that global warming will affect every aspect of life on Earth and that we should be prepared to counter its negative effects on plants, animals, and humans because without these entities the Earth would be a dreary place like the planet Mars.

CHAPTER THIRTY-THREE

Industrial Ecology: Toward an Interdisciplinary Approach to Sustainability

HENRY WADSWORTH LONGFELLOW, the 19th century poet, once suggested that people who make history leave "footprints on the sands of time." It is very likely that the footprints in the sand will be blown away, but it may resurface elsewhere in a different shape and significance.

Coining of the Word Ecology

The word ecology was coined in 1866 by the zoologist Ernst Haeckel, the "German Darwin." He defined it as "the relation of the animal both to its organic as well as its inorganic environment." The root of the word "oikos" is a Greek term meaning a home or a place to live.

Haeckel's message was that plants, insects, animals, and the environment in which they live cannot be understood in isolation from one another. Everything connects in a complex network of relationships and interdependence, but unfortunately ecology remained no more than a word on "the sands of time" until the 20th century when scientists began to put some substance behind the original concept as the globe was becoming industry's garbage can.

Our Stone-Age ancestors explored the world for safety from natural forces, other animals, and for sustenance. Humankind has always had insatiable curiosity about the world. At first, survival motives spurred human interest to find out what plants were edible, what streams were potable, what animals were safe to be around, etc. Now, the interest has shifted to a curiosity to know more about the marvels and mysterious of the world around and to protect the fragile environment and animals.

Agriculture the First Step to Civilization

About ten thousand years ago, humans quit as hunters and gatherers and became farmers. Agriculture has a place in the pantheon of human achievement for it contributed to the creation of civilization, a settled life, by people living together, forming a culture, producing art, literature, and architecture for generations to enjoy. Therefore, agriculture has served as the infrastructure for our early civilizations to take off and later to flower.

In the 1760s, the Industrial Revolution has moved humankind from subsistence level to a high standard of living at the cost of degrading or destroying the world in which we live. After two centuries of unabated pursuit of producing an abundance of goods to satisfy the glutinous appetite of the world population, humankind came to the realization in the 1970s that the environment was being exploited beyond repair for generation to come. Greenhouse effect, ozone hole, soil erosion, deforestation are but a few problems caused by human mismanagement. The phrase "Sustainable Development" was coined at a 1972 United Nations conference, which stuck as a term for justifying a range of policies.

It has been an accepted characterization that humankind is the most destructive animal on the planet Earth. Humans have exercised an unprecedented control over nature. As the population of the ancient world grew so did the need for food, clothing, shelter and even for luxuries too. Supply of necessities has always seemed to lag behind the growth in population by leaving a huge gap which created the proverbial divide between the poor and the rich. In an attempt to close the gap, humankind's curiosity turned into a quest to harness the resources of nature for the masses. Over the years, this move to push boundaries ushered in the Industrial Revolution by replacing human and animal power with machines. As a result, industrial waste was dumped at any convenient place around the world.

The practice of utilizing one resource from nature, however, caused the ripple effect to destroy other important elements in a given area. For example, cutting down trees for paper products, destroyed the habitat for birds and animals. In the 2000s, Canada's salmon population began to decline in certain rivers for no apparent reasons. This situation had become a conundrum for the experts. Upon close examination, it was found that industrial pollution was contaminating the water by making it unsafe for the fish to spawn.

The Ripple Effect in Nature

The realization of the ripple effect in nature has given a renewed interest in the concept of ecology as the idea was first proposed by the German zoologist years ago. It is not only a science or a practice but it is also a religion and a political weapon to wage war against industry for protecting the environment.

As we can see, two countervailing forces have emerged: industry is bent on pursuing efficiency in production to maximize revenues; on the other hand, ecology is crying out for a balance between the diverse elements in nature to prevent one element from overpowering another into extinction. Ecologist's passion is to preserve, while the business executive's primary goal is to produce what the customer needs or wants at a profit.

The only matchmaker to create a positive relation between the feuding industry leaders and ecologists is the Societal Marketing concept. This concept is the only business philosophy which, before producing any product to satisfy consumer needs or wants, would ask if such a move is consonant with the interest of society or the environment. If it is not, then the product is not produced. Ecologists would fall in love with industry if only industry would operate out of this philosophy. In this way, ecologists would cease battling with industry and settle down into a blissful matrimony of cooperation with the latter.

Industrial Ecology or "Indecology"

Young sciences have always had the tendency to branch out. For example, psychology branched out into social psychology, behavioral psychology, and industrial psychology. Likewise, ecology now has social ecology. Why not combine the two into "industrial ecology" or call it "Indecology" for short. In this way, executives as decision makers would be educated about the ecological consequences of their acts. Industry is defined as any branch of business, trade, or manufacture – all such enterprises taken collectively. "Ind" is an accepted abbreviation of industry. We can name the area anything we want, what is important is the content of the material to be learned.

Business students, our future executives, should be equipped with the knowledge to make decisions about sustainable developments which are friendly to the environment. Sustainability is the act of exploiting natural resources without destroying the ecological balance of a particular area through global resource depletion and environmental pollution.

We need scholarly and professional articles in the hope of improving our understanding of business and its practice locally and globally. Any descriptive article or study on the preservation of the environment in conjunction with any of the functional areas of business from this newly proposed interdisciplinary perspective, namely Industrial Ecology, or Indecology for short.

CHAPTER THIRTY-FOUR

Internationalism: The Imponderables of an Invasive Idea

DURING THE LAST one hundred years, humankind has generated ideas that shaped our world through such concepts as imperialism, liberationism, socialism, Marxism, industrialization, glasnost, and perestroika to mention a few. Paramount among these ideas is internationalism. Inherently, internationalism has a global perspective.

The Nature of Internationalism

Narrowly defined, internationalism refers to the recognition that all nations are bound together in a collective enterprise that calls for cooperative action. Due to modern methods of real time communication, the world has shrunk to a global village. As a result, internationalism has found fertile grounds to spread rapidly like a wild fire. Globalization fever, the offshoot of internationalism, can be likened to the California Gold Rush of 1845. Some prospectors struck gold; others panned nothing but gravel and sand. In its simplest form, though, it is a business idea that has been steadily shaping our world economies for better or worse.

Internationalism is the antithesis of protectionism, which is still practiced by some successful economies such as that of Japan's. Protectionism is the countervailing force against the notion of a borderless world. Yet despite the practice of protectionism, the majority of nations, China being a prime example of a late comer, seems to have embraced internationalism with open arms, eager to enjoy the fruits of capitalism and cooperation.

For most of the 20th century, Europe's destructive and decisive wars and the post-1945 Cold War between the capitalist West and the Communist Eastern Block, the energies of both sides had been mainly focused on militarism. The Western intelligencia were distracted from the greatest of international issues – the ever gaping economic disparity between rich and poor nations.

The dissolution of Communism in most Eastern Europe, beginning in 1989, brought the end of the Cold War and thus paved the road for the conduct of international relations. Meanwhile, the phenomenal advance of the Pacific Rim countries endowed with a mega population, a mega land, and a mega capital threw the established economic order into disarray.

The Advent of Globalization

In the 1980s, internationalism had metamorphosed into globalization. It started to emerge as an economic force, expanding Western as well as Pacific Rim economies. It consolidated itself during the economic uncertainty and recession of the 1990s, as multinational companies such as Sony, Siemens, and Ford located their plants and decision centers around the world simply to take advantage of low labor costs, to have access to cheap raw materials, and to obtain government grants.

During this time, advances in technology made it possible to transmit voluminous data electronically by either fax or e-mail in a few seconds. Improvements in transportation have also contributed to globalization. For example, fast air travel opportunities enabled companies to send goods from one end of the globe to another quickly and cheaply.

To free up international trade further, a number of agreements have been signed in 1990s. Among the most important ones was the North American Free Trade Agreement (NAFTA) of 1994, which has a program to establish free movement of goods and services among Canada, Mexico, and the United States by 2009. During this time another political initiative was The Maastricht Treaty of 1993. The main purpose was to set up the European Union, which also aimed to abolish trade barriers between its initial 12 member states.

Above all, after eight years of negotiations, the General Agreement on Tariffs and Trade (GATT) was signed in 1994 in Uruguay. The 125 participating nations agreed on a new code of conduct for international commerce. Its members also agreed to reduce tariffs through multilateral negotiations. All of the foregoing agreements have enhanced and encouraged international trade and business considerably.

The mirage of opportunities promised by globalization, however, represented a severe backlash for some countries. Certain countries were not getting economic parity with foreign companies. Michael Jordan, for example, makes more money from Nike annually than all of the Nike factory workers in Malaysia combined. Upon detecting the lack of economic playing field in globalization, some countries became apprehensive, leading to calls for renewed trade protection for trust is the currency of human exchange, be it social or business.

Japan, whose huge economic success is based on protectionist policies, face a tariff-free future with a great uncertainty. Western nations blame cheap Asian and Middle Eastern labor for job insecurity and wage freezes. Regardless of its fame or notoriety, globalization has already changed the rules of the economic game and can be as much a source of opportunity as a threat.

Globalization Draws Controversy

Currently, globalization is the single most controversial topic in internationalism. Myriad of articles and books have been written on its pros and cons. The pressures of international action nevertheless remain the same: is globalization a panacea to eradicate world ills and disparities between the haves and have-nots? Or is it God's curse to punish mankind by ushering in disease, pollution, over population in the economically challenged countries of the world, resulting in water, housing, food, employment, and education scarcity?

In spite of the attitudes held, between the two extremes of protectionism and globalization, the pendulum has swung toward the latter. A time may come when it would reverse its position and move toward the former. A shift in the direction of the pendulum may be plausible since nothing is constant in the business world except that of change. A sign of things to come is mirrored in the strategies of globalizing companies which are finding success as markets localize.

Pessimists predict the weakening of social systems in the third world countries, resulting in the collapse of sovereign states and the eventual disappearance of small ethnic groups. One African nation, for example, has a diverse population speaking about 27 different languages. Kids are learning English in schools; they are using it at the playgrounds, on the streets, and at home. The deep concern is that these traditional languages are bound to become extinct. When a language disappears, so does the ethnicity of the people. The optimists, on the other hand, foresee great economic improvements predicated on the free-market principles and economic interdependence between the rich and the poor nations.

Not only do I respect, but also encourage contributions of various perspectives on this important issue fraught with controversy. Certainly, internationalism stands imponderable despite its invasiveness in the landscape of world economies. Time will tell its strengths and weaknesses, its successes and failures, its promises and its disappointments.

Internationalism: The Imponderable Idea

While the experts debate the pros and cons of internationalism, we need a forum to represent both sides of this imponderable, yet ever increasingly invasive idea. The business sector has always recognized the emerging tidal force of internationalism. This consciousness has led to the inclusion of various mutations and dynamics of internationalism.

Moreover, the incidence of so many conferences on the theme of internationalism and globalization is testimony to our dedication to the proposition of providing readers with forums to discuss, debate, and evaluate this emerging idea of globalization which is the latest manifestation of internationalism.

The Western minds, the Eastern government officials are all weighing the perils and panaceas of internationalism which is defying any definitive evaluation of its impact on the economy, culture, and environment of the participant nations. Some hail globalization as the best idea ever happened to the world; others bemoan its disastrous effects.

Despite all the uproar and brouhaha, hoopla and the hooray, the idea of internationalism seems to be entrenching by sending its tap root deeper and deeper into the world business. The idea seems to have already become institutionalized, especially after the collapse of the Soviet Union. Cooperation without self-sacrifice, however, fails to cross-pollinate any critical progress among nations. To some newly independent nations, internationalism is fast becoming the opium of the people – the only way out of the vicious circle of poverty. I wish them all Godspeed!

CHAPTER THIRTY-FIVE

The Millennial Generation Mindset

AS A NEW generation emerges, historians, social commentators, and politicians among others begin to appraise the personality and the future productivity of those coming of age within a society.

Recently, 60 Minutes (CBC News), Bill Moyer's Journal, The Pew Research Center, and many academic articles have their take on the new kids on the block. Some of the comments made on their behalf have been positive; while others have not been so favorable, even bordering on being very negative.

The Millennials

Those who hold positive attitude toward the Millennials maintain that these American teens and twenty-somethings who have made the passage into adulthood at the start of the new millennium – have begun to exhibit confident, self-expressive, liberal, upbeat and open to change characteristics. They are more ethnically and racially diverse than older adults. Their religiosity is frail, less likely to have served in the military, and are on track to become the most educated and tech savvy generation in American history.

They are history's first "always connected" generation. Steeped in digital technology and social media, they treat their multi-tasking hand-held gadgets with reverence. More than eight-in-ten say they sleep with a cell phone glowing by the bed, poised to spew phone calls, emails, texts, songs, news, videos, and games. Nearly two-thirds admit to over texting.

Those who hold negative attitudes toward the Millennials, levy the charge that they are the first generation in the history of the United States who expect not to do better than their parents. Hence, they have been characterized as being less ambitious, lazy, whiners, devoid of leadership qualities, and electronic gadget addicts.

Every coin has two sides, so does the personality of the Millennials. America's newest generation is in the middle of this coming-of-age phase of its life cycle. It is fast maturing; its oldest members are approaching age 35 and its youngest are approaching adolescence.

There are over 80 million of them now, born between 1980 and 1995, and they are rapidly taking over jobs and positions from the baby boomers who are now pushing 70.

The Pessimism about the Millennials Traits

How can we explain the expressed qualms and pessimism about the Millennials traits? Well, it is true that their entry into careers and first jobs has been badly set back by the Great Recession.

The workplace has become a psychological battlefield and the "Millennials" have the upper hand, because they are tech savvy, with every gadget imaginable almost as though they have become joined twins. They multitask, talk, walk, listen and type, and text every moment of the day. Their priorities seem to be simple: they come first. However, because of the persistent recession, many Millennials have remained jobless and hence face economic hardship since they owe the government for the money borrowed for their education.

Since September 2008, the world press has begun to monitor events to determine the effects of the financial crisis that began in the USA on various international economies. It is amazing how fast the financial crisis from the United States extended and transformed into an economic crisis worldwide. All foreign national economies had become affected to a certain extent.

Likewise, it is unbelievable how a crisis generated another crisis. For example, at the end of 2008, business analysts launched pessimistic forecasts regarding the evolution of national economies in 2009, but they considered scarcely the issue of social crisis generated by the economic crisis. The conference entitled "Crisis of Confidence" had exposed many culprits in creating the massive financial malaise.

"The Recession and the Economy of Fear" conference, sponsored by the University of Pennsylvania's Department of Psychiatry and the Psychoanalytic Center, emphasized the following view: the individual's ". . . emotion not only led America into the present economic crisis but it could also keep it there."

Loss of Confidence in the Economic System

David M. Sachs, training and supervising analyst at Psychoanalytic Center of Philadelphia, stated that "the economic crisis is not one of concern but one of confidence". The emotional response of consumers to the effects of the financial crisis determined the decrease of their confidence in trademarks, organizations, governments, etc.

In other words, the negative emotional response had determined the appearance of confidence crisis. The negative economic evolutions and the decrease of consumers' confidence implied the restriction of consumption. Consumers began considering savings as a reaction to the uncertainness of their present existence. As a consequence, this feeling of fear and uncertainly of the future caused markets to contract.

Thus, emotions of fear contributed to economic crisis, and confidence crisis contributed to a sharpening of economic crisis. The resultant effect was an overall decrease of consumer confidence in trademarks, companies, fields of activity, governments and the near future.

The effect of economic crisis on one's confidence is consistent with some of the findings of studies on the negative effect of weather on the individual's mood, productivity, and frequency of emotional stability. Millennials present mindset could very well be a reflection of the nagging hard times determined to undermine world optimism to yearn for economic progress of the past century.

Perhaps, when the world economy climate improves, when we enter an area of prosperity full of "sunshine and blue skies" hopes, so will the Millennials change their attitudes toward the future. After all, the Millennials are human like the rest of us and thus they are susceptible to changes in their social, economic, and political environments. I personally wish our young Millennials Godspeed in every aspect of their endeavors!

CHAPTER THIRTY-SIX

Next to Dogs, Man's Best Friends Are Dying: Vanishing Bees and Businessmen

THE LAST 200 years, distinguished as they have been by the advance of technology and industrialization, have been an era of unprecedented scope and rapidity of social and economic changes in human history. Although humankind has made many strides in various sciences, but when it comes to the disappearance of the humble honeybee, biologists have no explanations but conjectures proposed as hypotheses.

Colony Collapse Disorder

In 2006, a baffling phenomenon fell upon honeybee hives across the world. Without leaving a trace, millions of honeybees disappeared from their hives, never again to return to their "honey factories" (i.e., nests). The incident has been called Colony Collapse Disorder (CCD) by biologists. From time immemorial, bees have been regarded as one of man's best friends. They are precious pollinators of fruits and vegetables. Scientists are scrambling to find answers to as why the bees were vanishing or dying in large numbers. This has become an urgent global mystery.

Had Charles Darwin been around, he would have settled the mystery as the work of evolution in the sense that bees have declared independence from living and working in a regimented colony like slaves in contrast to the religious belief that the great diversity of life on

Earth and the web of interdependence between different species are powerful evidence for a Creator.

Causes for the Disappearance of Our Friends

The culprits in the vanishing bees range from psychological causes such as stress for commercial bees are hauled by truck, rail, ship, or by air from orchard to orchard, from country to country to pollinate, to physiological causes such as a virus called IAPV which was discovered in Israel in 2004. Other plausible causes include over-use of pesticides, parasites, poor nutrition, pests like varroa mites which trigger a virus, etc. United States imports bees from Australia and New Zealand five percent of it just needed to pollinate the almond trees alone. Hauling of bees from such a long distance creates many problems for the bees.

The Implications of the Vanishing Bees

The implications of the vanishing bees are far reaching. Billions of dollars of crops are at risk and threaten our food supply. Therefore, the effect of CCD on world agriculture is enormous. Since bee population decline has been observed all over the world, this has become a global problem of great magnitude. For example, in the winter of 2006/2007, the United States lost a quarter of the nation's 2.4 million bee colonies; France lost immeasurable numbers of bees; Spain, because they have the highest number of commercial beekeepers in Europe, suffered massive losses; England claims that their beekeepers lost two thirds of their stock; and China suffered great losses in some regions where time-consuming, tedious, and costly hand pollution was carried out in some fruit orchards.

From business standpoint, the loss is projected to be between $8 billion to $12 billion on U.S. agricultural economy alone. The role of the honeybee transcends the mere making of honey. This noble creature pollinates about one-third of crop species in the U.S. They pollinate about 100 flowering food crops including apples, nuts, broccoli, avocados, soybeans, asparagus, celery, squash and cucumbers, citrus fruit, peaches, kiwi, cherries, blueberries, cranberries, strawberries, cantaloupe, melons, as well as animal-feed crops, such as the clover that is fed to dairy cows. The list also includes the billion-dollar industry of ornamental plants sold for landscaping. Essentially, all flowering plants for food or ornament need bees to survive.

The world population is threatened with more widespread famine to add to the existing problems of shortage of food due to dust bowls in Africa and the environmental changes precipitated by global warming. So far the study of the disappearing honeybees has been carried out by biologists. In fact, the topic should also be included into the domain of business research and writing. Business professionals and academicians have also to engage in research to size up the effect of the loss of bees on the economy, on the variety of foods placed on the consumers table, and the price of these fruits and vegetables owning to the less efficient pollinators such as depending on other insects and birds to do the great job of the honeybee. Nothing would replace the bees; it would be like sending out a house cat to do the big job of a tireless Bengal tiger.

The Ripple Effects from the Vanishing Friends

We are all in this together. The ripple effects from the vanishing bees touch us all regardless of our social, economic, geographic, and professional backgrounds. It is incumbent on all physical, social, and behavioral disciplines to work hard toward the return of man's "other" best friends to our yards, orchards, and fields for the silence of bees would cause immeasurable damage to human and animal sustainable standard of living. We need to study the implications of the vanishing bees from various business angels. While dogs are considered to be man's best friend, without bees man and his sidekick the dog would go hungry and the world would turn grey.

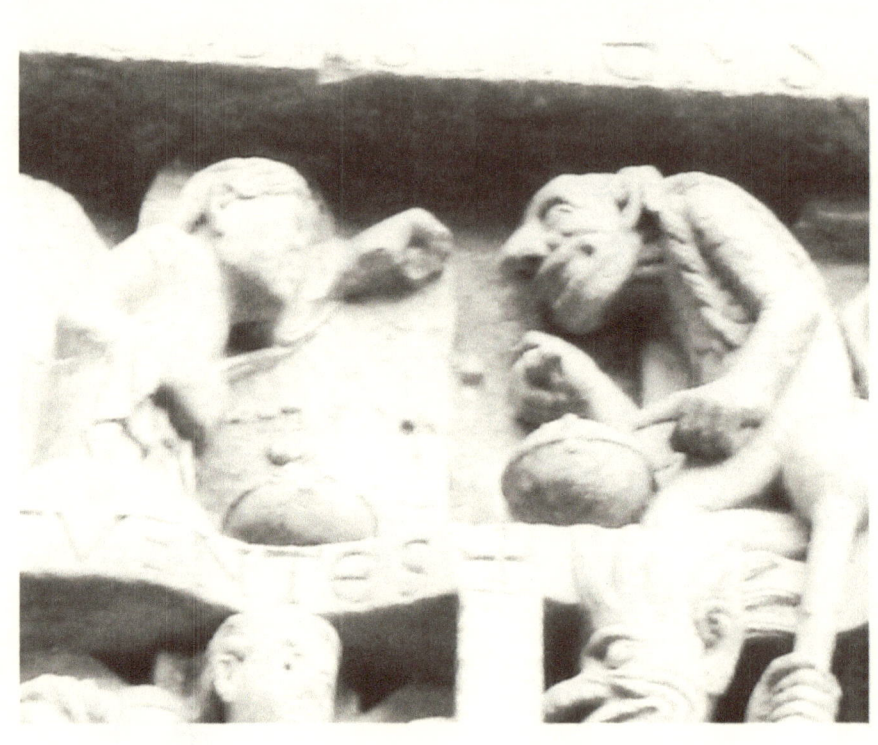

CHAPTER THIRTY-SEVEN

POLITICAL Lobbying: Legitimizing the LETHAL TOOL of Marketing

AGAINST THE BACKDROP of increasing world economic and business problems due to globalization and political conflict proliferations, the lobbying practice as a lucrative industry is growing in leaps and bounds. Government lobbying has become a growth industry. Corporations and nations are increasingly resorting to lobbying services to protect their interests when decisions are made in domestic and foreign relations diplomacy.

Despite the exponential growth of lobbying services at the K Street in Washington, D.C. and the European lobbying in Brussels, Belgium, little is known about its secrets of organized operation, its relation to marketing, and its drastic need for regulation to "ethicize" its process in providing decision makers with valid information.

Lobbying in the Days Past

The profession of lobbying goes back over two hundred years. Currently, there are now more than 27, 000 registered lobbyists in Washington, D.C. and 15,000 lobbyists in Brussels petitioning on behalf of business, labor, the environment, education, abortion rights, the elderly, the poor, ethnic groups, and foreign nations. Almost all of foreign nations jockey for military, economic aid, and other policies and decisions to enhance their positions.

In an effort to get decisions made in the interest of their clients, some lobbyists resort to shady and shameful methods. In fact, lobbying is the legitimate, but neglected child of marketing, which, like a wild Tarzan, operates almost freely in the political jungle of Washington, D.C. through campaign fund contributions, appointment of individuals in key positions, and through the ministry of fear. Certain lobby organizations have wielded so much power through unethical procedures that they have held Congress and the White House hostage. Presidents are elected to office and presidents are toppled by the lobbying machine.

Despite its importance, the marketing literature lacks a model that delineates its place in the overall scheme of marketing, shows how it works, demonstrates how it should come under the legislations which basically regulate marketing, and, most importantly, synthesizes a new code of ethics for corporate lobbying based on marketing and political lobbying codes of ethics.

The recent Jack Abramoff's scandal on the Capital Hill and the resignation of Senator Robert Nye have ushered in a new urgency to deal effectively with the unethical conduct in business and government. The pressing need for the search of ethics in lobbying for business and government has never been greater than at the present time. Ethicizing business transactions has become an absolute imperative for a civil society. With the advent of Enron's debacle, lobbying reform has become a dire necessity for a healthy business and society. Abramoff's unethical conduct inspired the Legislative Transparency and Accountability Act of 2006, which still contains many loopholes for deception and corruption.

The current lobbying practices will take center stage in the next presidential campaign issues for its lack of ethics. The clamor for Lobbying reform will definitely not subside until presidential candidates promise to take stringent measures against the highway to deceptive lobbying practices.

The Need for a Model of Lobbying

Any proposed viable model should show the relationship of lobbying to its parent marketing, the legal jurisdiction under which it should also come, and the newly synthesized code of ethics will be integrated in the teaching of ethics primarily in the Principles of Marketing and Marketing Management courses. The need for a pedagogical framework is sorely needed especially in the capstone courses of marketing in which case analysis method is highly emphasized.

Based on the experiences of some socially responsible business executives and drawing upon a number of highly publicized cases of breach of ethical conduct along with their consequences, some recommendations for ethical conduct should also be summarized and presented to serve as a guideline for proper business behavior and decision making.

The basic requirement is to construct a comprehensive model which shows how lobbying is related to marketing and how it functions in the business world. After having delineated the practice of lobbying in marketing, lobbying should show how it will fall under the legislations enacted to regulate marketing. The proposed model should also be used as a framework to guide empirical studies by other faculty members. A thorough review of the literature failed to disclose of a paradigm which describes and explains the construct of lobbying theory and practice to advance research and writing in this critical area which has enormous influence on business and society.

Business Corporations are Waging War

Business corporations are increasingly waging war on one another through their mighty lobbyists. To foster ethical leadership, there is a great need to show how lobbying is related to marketing and how it works within a democratic society despite its maligned image due to some unethical lobbyists who have brought discredit to the profession. Ethical lobbying practices contribute positively to business and society, and to the democratic processes of the government. After all, only ethical captains of business and industry deserve to be enshrined as outstanding leaders of our society and as role models for students, our future executives, business leaders, and stewards of industry.

CHAPTER THIRTY-EIGHT

The Women's Millennium in a Man's World

THE STRUGGLES TO gain social, economic, and political equality with men are long sagas in the annals of women's history. So long as historical writing has been in the sphere of men's influence, little coverage has been devoted to record the feminism movement in the world.

The Modern Movement of Feminism

The modern movement of feminism, however, is usually traced to the late 18th century, to the formation of women's political clubs in Paris in the wake of the French Revolution. Furthermore, this movement is also traced to the nascent campaign for female suffrage ushered in by the 1792 publication of ***Vindication of the Rights of Women*** by Mary Wollstonecraft, an English writer (1759-1797).

The crusade of the feminists in the 19th century centered on gaining for women the right to vote, property rights in marriage, and the right to be educated. Twentieth century feminism in the West, however, arising strongly in the United States in the 1960s, has targeted male patriarchy in all its aspects. Prominent among the issues attacked has been gender discrimination in a wide variety of fields beyond the political and economic parameters. A few examples of the issues attacked included linguistic studies, psychoanalysis, history, and many others. Feminism has become both a tool of analysis and a practical program to ameliorate women's lot in a predominantly man's world since time immemorial.

Among the women's achievements in varying degrees in the world have been legal recognition of their rights to equal pay with men for equal work done, to equal division of property in divorce, and to freedom in abortion decisions. The struggle continues to dismantle the Victorian mythic pedestals upon which men had enshrined themselves for too long.

A Milestone in the History of Feminism

In 1988, Working Woman magazine surpassed all other North American business magazines in circulation, including the well entrenched Fortune, Forbes, and Business Week. For 350 years, men outnumbered women on college campuses. Women have been put down in the United States as being inferior to men in many areas. Just a century ago, the president of Harvard University, Charles W. Elliot, refused to admit women because he feared they would waste the precious resources of his school.

In the last 30 years, however, women have imperceptibly gained the first place in education. In 1991, for the first time ever, female students outnumbered their male counterparts in colleges and universities. In the year 2000, studies have shown that female students had begun to outperform male students in education. Since 1971, the number of women earning degrees in the field of business has grown astronomically. Volvo exhibited the first automobile ever for women designed by women on display in Geneva in March 2004. On Saturday, March 6, 2004 history was made in Marathon competition. A 49-year-old woman for the first time won the Los Angeles Marathon. What a feat of achievement, despite the twenty minute of lead-time granted to women competitors. As in business, glass ceilings are also being shattered in the landscape of the corporate world. The list of accomplishments through women's silent revolution goes on and on to the point where one can dub the zeitgeist as being "the women's millennium."

The Case of the Journal of American Academy of Business

The Journal of American Academy of Business, Cambridge (JAABC) is proud to indicate that a good number of their contributors have been female professionals and academics as shown in Table 1.

Table 1

The Journal of American Academy of Business, Cambridge Article Contributors' Analysis*					
	MALE	FEMALE	TOTAL	Male %	Female %
Vol. 1, No. 1, September 2001	23M	4F	27	85%	15%
Vol. 1, No. 2, March 2002	33M	10F	43	77%	23%
Vol.2, No.1, September 2002	73M	8F	81	90%	10%
Vol.2, No. 2, March 2003	72M	5F	77	94%	6%
Vol.3, No.1&2, September 2003	81M	11F	92	88%	12%
Vol.4, No.1 & 2, March 2004	97M	12F	109	89%	11%
Totals	379M	50F	429	88%	12%

*Data were compiled by Dr. Turan Senguder

Although male contributors outnumber their female counterparts, those in the journal publishing business would recognize that the female proportions in the above table as being above average if compared to other similar publications. Even without comparison to other publications, if only absolute numbers were considered, JAABC would emerge as a journal attracting a substantial number of female contributors.

Given the opportunity, women are proving themselves capable on all fronts of human endeavor. As we have seen, they are even excelling in certain areas such as in education. JAABC salutes women's achievements and their determination to seek fairness and equity, which have been elusive for them for so many centuries.

Humankind has only seen the tip of the
iceberg of nature's deep mysteries.
Science is the key to unlock the hidden
secrets of the environment in which we live.

BOOKS AND ARTICLES FOR FURTHER READING

List of Books

Allenby, Brade R., and Graedel, T. E. (2002). Industrial Ecology. Prentice Hall, 363 pages.

Barnett, Christopher Brendan and Daniel G. Maxwell (2005). Food Aid After Fifty Years: Recasting Its Role. New York. Routledge, 314 pages.

Berinstein, Paula (2003). Business Statistics on the Web: Find Them Fast-at Little Or No Cost. CyberAge Books, 244 pages.

Collins (2006). Fragile Earth: Views of a Changing World. Nook Book

DK Publishing (2013). Ideas That Changed the World. New York, DK Publishing

DeSalle, Rob, and Yudell Michael (2002). The Genomic Revolution: Unveiling the Unity of Life. Joseph Henry Press, with the American Museum of Natural History, 249 pages.

Drezner, Daniel W. (2007). All Politics is Global: Explaining International Regulatory Regimes. Princeton University Press, 234 pages.

Ethridge, Marcus E. (2012). Politics in a Changing World, 6th Edtion. Cengage Learning

Ferrell, O.C. (2008). Business: A Changing World, 7th Edition. The McGraw-

Hill Companies Friedman, George (2012). The Next Decade Empire and Republic in a Changing World. Nook Book

Kabber, Naila (2003). Gender Mainstreaming in Poverty Eradication and the Millennium Development Goals: A Handbook for Policy-makers and Other Stakeholders. Canada. Commonwealth Secretariat, 245 pages.

Komblum, William (2011). Sociology in a Changing World, 9th Edition. Cengage Learning

Morgan, Robin (2003). Sisterhood Is Forever: The Women's Anthology for a New Millennium. New York. Washington Square Press, 576 pages.

Moller, J. Orstrom (2000). The End of Internationalism: Or World Governance. Greenwood Publishing Group, 205 pages.

Omae, Kenichi (2005). The Next Global Stage: Challenges and Opportunities in Our Business World. Wharton School Publication, 282 pages.

Ratner, Mark A. and Ratner, Daniel (2003). Nanotechnology: A Gentle Introduction to the Next Big Idea. Prentice Hall PTR, 188 pages.

Redclift, Michael (2005). Sustainability: Critical Concepts in the Social Sciences. Taylor & Francis US, 1576 pages.

Rowntree, Lester (2013). Globalization and Diversity: Geography of a Changing World, 4th Edition. Prentice Hall

Rugman, Alan M. and Verbeke, Alain (1990). Global Corporate Strategy and Trade Policy. Routledge, 168 pages.

Winter, Beborahh Du Nann., and Koger, Susan M. (2004). The Psychology of Environmental Problems. Routledge, 287 pages.

Woodstock Theological Center (2002). The Ethics of Lobbying: Organized Interests, Political Power, and the Common Good. Georgetown University Press, 97 pages.

Z., Can (2012). Changing Worlds. Nook Book

List of Articles:

Allenby, Braden R. and Rejeski, Dave (2008). The Industrial Ecology of Emerging Technologies. Published Online: Sep 3 2008 6:39PM in Journal of Industrial Ecology, Yale University (p 267-269). www.interscience. wiley.com/jpages/1088-1980

American Solar Energy Society (2005). Solar Today. American Solar Energy Society, an original report from the University of Michigan, Digitized Jan 18, 2008.

Ashraf, Tariq (2004). Information technology and public policy: a socio-human profile of Indian digital revolution. Published by Elsevier in The International Information & Library Review, India, September 1, 2004, Volume 36, Issue 4, December 2004, Pages 309-318

Attfield, Judith (2003). What Does History Have to Do With It? Feminism and Design History. Journal Design History 2003 (A CSU Long Beach Magazine) Volume 16, Number 1 Pp. 77-87

Borger, Julian (2006). Lobbyist to reveal all in Congress bribes scandal in Washington. The Guardian, on Wednesday January 4 2006

Chandler, David (2008). On the front lines of the genomic revolution, Office on February 6, 2008 in MIT News (Massachusetts institute of technology).

Dzenis, Yuris (2004). Spinning Continuous Fibers for Nanotechnology. Science 25 June 2004: Vol. 304. no. 5679, pp. 1917-1919

Ehrenfeld, John and Gertler, Nicholas(2008). Industrial Ecology in Practice: The Evolution of Interdependence at Kalundborg. Journal of Industrial Ecology Volume 1 Issue 1, Pages 67-79 Published Online: 8Feb, 2008 ©2008 Yale University

Howard, A., & Wellins, R. S. (2008). Global leadership forecast 2008-2009: overcoming the shortfalls in developing leaders. Pittsburgh, PA: DevelopmentDimensions International.

Galbreath, Jeremy (2005). Corporate social responsibility strategy: strategic options, global considerations. Curtin University of Technology, Perth, Western Australia in 2005, See website xtra/&emeraldinsight.com

Jones-Evans, Dylan (2008). Social Entrepreneurship – Getting support from Government. University of Wales on June 16, 2008

Martin, Kevin (2007). The looming leadership void: identifying, developing, and retaining your top talent. Boston, MA: Aberdeen Group. dylanje.blogspot.com/2008/06

McCain, John (2008). Self-Confidence on Ethics Poses Its Own Risk. New York Times on February 21, 2008

Nevo, Eviatar (2008). Evolution of genome – phenome diversity under environmental stress, Journal PNAS, Israel October 21, 2008, 105 (42)

Powell, Colin L. (2003). American. Internationalism, published by U.S. Foreign Agenda. An Electronic Journal of the U.S. Department of State. Volume 8, Number 1. August 2003.

Schwab, Klaus (2008). Global Corporate Citizenship Working With Governments and Civil Society. Published by Council in Foreign Relations in January/February 2008

Seeman, Corey (2008). Digital and Microform (microfilm/microfiche) technologies and . . . . Microform & Imaging Review international journal, University of Michigan from 05/01/2008-06/01/2008, see website: librarywriting.blogspot.com/2008_05_01_archive.html

Sharma, Pramodita (2005). Trends and directions in the development of a strategic management theory of the family firm. Publication: Entrepreneurship: Theory and Practice on September 1 2005, India.

Solar Energy News and Articles (2008). Hope for a Solar Future – U.S. Reverses Its Moratorium on Solar Energy Research. Published by www.naturalnews.com/solar_energy.html on solar energy news and articles . . . . 10/7/2008 – (Natural News) On June 27th . . .

Stern, Nicholas and Taylor, Chris (2007). Economics: Climate Change: Risk, Ethics, and the Stern Review. Published in Science 13 July 2007: Vol. 317. no. 5835, pp. 203-204

Valencia, Albert (2008). Discussion of Rapadas: Transmission of Violence: Parallels to Latinos in the United States. California State University, Fresno, Australian Academic Press.

Walby, Sylvia (2005). Gender Mainstreaming: Productive Tensions in Theory an Practice. Social Politics (A CSU Long Beach Magazine) Volume 12, Number 3 Pp. 321-343 November 8, 2005

SUBJECT INDEX

D

corporations, 227
vinicultural history, 137

W

Walter Raleigh's ideal,
    176
war in Afghanistan, 224
water purification, 66,
    195
water purification bottles,
    66
Weapons of War, 308
Web with the Internet,
    109
Wegener's concept, 14
Western paradise, 315
Western Union, 311
West is a "paradise," 297
white goat, 56
White House, 35, 38, 123,
    425

wildebeests' migration,
    291
women's millennium,
    431, 434
WorldCom, 78
World Health
    Organization, 350
world Inuit organization,
    272
World Trade Center in
    New York, 74
World Wide Web
    (WWW), 107

X

xenogenesis, 133
xenotransplantation,
    133-34

NAME INDEX

A

Abramoff, Jack, 425
Addams, Jane, 190
Allenby, Braden R., 439
Allenby, Brad R., 442
Archimedes, 145-46, 258
Aristotle, 88-89, 205,
    258, 335
Ashraf, Tariq, 442
Attfield, Judith, 443

B

Barnett, Christopher
    Brendan, 439

Bell, Daniel, 281-83, 286
Berg, Paul, 130
Berinstein, Paula Collins,
    439
Berners-Lee, Tim, 108,
    112
Blanc, Louis, 232
Bojarsky, David, 11
Bonaparte, Napoleon, 44,
    50
Bond, James, 307
Borger, Julian, 443
Boyer, Herbert, 131
Bush, George W., 222
Bygren, Lars Olov, 345

www.ingramcontent.com/pod-product-compliance
Lightning Source LLC
Chambersburg PA
CBHW020720180526
45163CB00001B/41